Systems

Principles and Practices

Cary R. Spitzer

Second Edition

First Edition Published in 1987 by Prentice Hall
Second Edition Published in 1993 by McGraw-Hill, Inc.

Copyright © 2000 by Cary R. Spitzer
"All rights reserved. No part of this book may be reproduced, stored in a retrieval system, or transmitted by any form or by any means, electronic, mechanical, photocopying, recording, or otherwise, except as may be expressly permitted by the applicable copyright statutes or in writing by the publisher."

ISBN: 1-930665-12-1

Library of Congress Card Number: 00-108987

THE BLACKBURN PRESS
P. O. Box 287
Caldwell, New Jersey   07006
www.BlackburnPress.com
973-228-7077

To my wife, Laura, and my son, Danny, in appreciation of their continuing love, understanding, patience, and inspiration to write this book.

To my colleagues who are designing, building, and installing digital avionics so that aviation may achieve its full potential.

Contents

Preface xi

Chapter 1. Establishing the Avionics System Requirements 1

The Aircraft and Its Mission Drive the Avionics System Design 1
Mission Analysis Techniques 6
A Look at the Requirements of MIL-F-9490 13
Concluding Remarks 17
References 18
Bibliography 18

Chapter 2. Avionics Systems Essentials I: Data Buses 19

Data Buses Using MIL-STD-1553 20
DOD-STD-1773 Is the Fiber Optic Companion to MIL-STD-1553 27
High-Speed Data Bus 29
Data Buses Using ARINC 429 31
Data Buses Using ARINC 629 37
ARINC 659 Is the Civil Avionics Backplane Data Bus 43
References 46
Bibliography 47

Chapter 3. Avionics Systems Essentials II: Crew Interfaces 49

CRTs Are Versatile Display Devices 50
Luminous Flat Panels 54
A Look at Liquid Crystal Displays 56
Comparing the Display Media 58
Multifunction Keyboards Save Space, Offer Flexibility 60
Head-Up Displays Offer Innovative Options 61
Head Level Displays Offer a New Option 64
Helmet-Mounted Displays 65
Night Vision Goggles 66
Laying Out the Cockpit 67
Voice Interactive Systems Are Another Way to Communicate 71
From Esoteric to Exoteric: Display Terms 75
References 75

Chapter 4. Avionics Systems Essentials III: Power — 77

- MIL-STD-704 Applies to Military Aircraft — 78
- DO-160 Prescribes Civil Aircraft Electrical Power Quality — 81
- Comparing the Military and Civil Requirements — 85
- Tips on Power System Design — 85
- References — 90

Chapter 5. Fault Tolerance — 91

- Fault-Tolerant Hardware and Systems — 92
- Fault-Tolerant Software — 99
- References — 100

Chapter 6. Maintainability and Reliability — 103

- Maintenance: Who, When, Where? — 104
- Designing for Easy Maintenance — 105
- BITE and CFDS Are Big Helps — 107
- Automatic Test Equipment Speeds Maintenance — 111
- ATLAS: A Giant of a Test Language — 112
- Reliability Is Equal Parts of Art and Science — 114
- References — 117
- Bibliography — 117

Chapter 7. Architectures — 119

- Fundamentals of Architectures — 119
- Examples of Modern Architectures — 121
- References — 138
- Bibliography — 138

Chapter 8. Packaging and Fitting the System into the Aircraft — 139

- Civil Packaging Standards — 139
- Military Packaging Standards — 142
- Keeping the Avionics Cool — 145
- Specifying Interfaces with the Aircraft — 148
- References — 149

Chapter 9. Hardware Assessment and Validation — 151

- Tips on Qualitatively Evaluating Systems — 152
- Major Regulations for Civil Avionics Certification — 154
- Fault Tree Analysis Is a Proven and Accepted Technique — 157
- Failure Modes and Effects Analysis — 162
- Criticality and Damage Modes and Effects Analyses — 169
- DO-160 Test Requirements — 173
- MIL-STD-810 Test Requirements — 176
- Electromagnetic Interference — 178
- Quantitatively Evaluating Systems Designs — 184

Examining the "ilities"	187
References	189
Bibliography	189

Chapter 10. Software Design, Assessment, and Validation — 191

The Essentials of Software Development	192
Some Helpful CASE Tools	195
What DO-178 Requires	200
What DOD-STD-2167 Requires	205
What DOD-STD-2168 Requires	211
Using Ada	211
MIL-STD-1750 Instruction Set Architecture	216
References	218
Bibliography	219

Chapter 11. Figuring the Costs of Avionics — 221

Some Underlying Principles	221
Life Cycle Costs for Civil Avionics	223
Cash Flow Analysis for Civil Avionics	226
Life Cycle Costs for Military Avionics	227
Software Costs	231
Establishing the Spares Level	234
References	240

Glossary	241
Appendix A. Environmental Testing	249
Appendix B. Software Documentation Description	261
Appendix C. Document Ordering Addresses	271

Index 273

Preface

The 1990s are the decade when digital avionics becomes a full-fledged, essential member of the aviation team. The role of avionics in aviation has continually expanded from the earliest primitive radios through the first generation of bulky, unreliable airborne radars to today when some aircraft cannot fly without avionics and virtually every aircraft relies on avionics to perform critical functions.

This ever-expanding role for avionics is the result of hundreds of technology pulls and requirements pushes. In engineering parlance, there is a knee in the curve of avionics applications that matches the evolution of digital microelectronics. As quickly as the avionics technologist developed new capabilities, built on the exploding world of microelectronics and software, the aircraft operators were calling on the avionics to do more. A new phrase, *digital avionics*, was coined and became a familiar term, at least to the cognoscenti. Digital avionics made possible incredible new aircraft capabilities, and avionics became mandatory for safe, efficient flight. In the future, the growth in digital avionics is expected to match, or perhaps exceed, that predicted for the microelectronics that make the avionics possible.

Over the last decade or so, concepts, methods, requirements, and specifications for designing, building, testing, and installing digital avionics have been developed and published in a plethora of documents. A measure of the maturity of any technology is when such diverse material can be captured and synthesized into a single volume. This book marks that measure for digital avionics by presenting the best and most important material from many sources, along with essential and practical tips on how it can be applied. Emphasis is placed on effective, efficient digital avionics systems rather than on the system components.

On these pages, the practicing engineer—the person on the firing line and the one who catches it when things go wrong—will find techniques, methods, and sometimes even a little black art that has worked in the past. There is guidance for every phase of a digital avionics system, from conception through installation to maintenance. Other readers, especially managers, will find this book useful in broadening their understanding of the issues in digital avionics and in developing a technical rapport with avionics technologists.

Even though this book focuses on digital avionics systems, many of the principles and concepts presented here can be applied with equal success to other electronic systems. Thus, the book may have a larger application than suggested by the title. The emphasis in this book is on flight controls and vehicle management avionics because of their unique, especially demanding role; however, the material presented is equally relevant to all other avionics.

What This Book Has to Offer

This book is designed to lead the reader through a logical flow in designing, building, evaluating, installing, and maintaining a digital avionics system. It begins with defining the requirements and flows from there through the system building blocks; to designing, packaging, and evaluating the system; and, finally, to a discussion of the costs.

An effective digital avionics system begins by understanding the role and mission of the aircraft. With that understanding, you can then establish the role for digital avionics. In Chapter 1 there is a series of topics to be addressed in defining the avionics requirements and some particularly demanding requirements for flight control systems.

Every digital avionics system, whether simple or sophisticated, has the common elements of data buses, displays, and power. The next several chapters look at each of these elements on an individual basis.

The most widely used military and civil data bus protocols are mature and well defined, and their use is mandated in all modern avionics systems designs. Chapter 2 summarizes the major data buses and also takes a look at some emerging bus concepts.

Since we cannot build aircraft that can operate very effectively without someone in the control loop, we must have a crew (or, in the case of an unmanned aircraft, a ground-based operator) and must let them know where they are and what the aircraft is doing. Seat-of-the-pants flying and electromechanical indicators are fading into oblivion. Chapter 3 contains information on display media and input-output devices, including voice. Using these devices, you can design a cockpit that the crew will enjoy using as much as you enjoyed designing it.

As avionics perform more important, even flight critical, functions, uninterrupted power is a must—any loss of power is catastrophic. Chapter 4 examines power system performance requirements and offers helpful guidance on designing this essential element.

Fault tolerance is necessary to achieve the extraordinary levels of reliability required of flight critical avionics. Simply throwing redundancy at the reliability problem will not solve it. Chapter 5 introduces the concepts of hardware and software fault tolerance, which are the only viable means to achieve flawless performance for critical functions.

Maintainability and reliability and, consequently, availability are the most important factors to aircraft operators—civil or military—after the avionics enters service. When a piece of avionics fails, the flight line mechanic must be able to replace or fix it quickly and reliably. Chapter 6 offers some insights into reliability and presents some proven maintainability concepts.

Once the avionics requirements are established and the basic building blocks and design essentials are understood, you can begin designing, building, and testing the system. In Chapter 7 the basics of architectures are introduced, and, in perhaps what is one of the most useful parts of the book, examples of modern digital avionics systems architectures are presented. The examples include the USAF Pave Pillar, the Beechcraft Starship, the Airbus A320, and the Boeing B-777 Airplane Information Management System.

An essential part of any avionics system design is fitting the equipment into the aircraft. Where can you put the black boxes? Can anyone get to them to work on them? Can you keep them cool? Chapter 8 explores these issues.

One of the major issues in digital avionics systems is the assessment and validation of the hardware, especially when the probabilities of system failure are expected to lie in the range of 10^{-5} to 10^{-9} per flight hour or even less? How fault tolerant is the system? Is the level of redundancy sufficient? Can the avionics survive and operate in the anticipated environment, especially the electromagnetic environment induced by lightning and high-intensity radio frequency transmissions? Chapter 9 helps to answer these questions. Although the answers to these questions depend on your own unique situation, you will find the right issues to address and the right steps to take in addressing them.

Software is just as important as hardware in a digital avionics system and requires the same careful assessment and validation. Chapter 10 discusses the unique nature of software development and assessment and validation. The chapter also looks at Ada, a high-order language that has been mandated for use on most military avionics and is the preferred language for many civil applications.

All of the avionics on the aircraft, even that required for flight critical functions, must pay their way through improved aircraft performance or economy of operation. Chapter 11 looks at the life cycle cost of avionics and offers detailed methods for computing these costs. One of the most popular methods for estimating software development costs is presented.

There is a glossary of commonly used avionics terms, and the three appendixes contain additional information. Appendix A provides more details on the environmental tests discussed in Chapter 9. Appendix B describes the documentation required as part of the software development processes reviewed in Chapter 10. Appendix C lists document ordering addresses.

Cary R. Spitzer

Digital Avionics Systems

Chapter 1

Establishing the Avionics System Requirements

Every digital avionics system that performs as promised has its roots in a clear definition of the system performance requirements. The designer cannot hope to build a system that will satisfy the customer if the design process begins with a fuzzy, incomplete concept of what the system is expected to do. This opening chapter discusses the development of system performance requirements based on a clear understanding of the needs of the customer.

The host aircraft and its mission are the top-level inputs to the definition of the avionics system requirements. The designer must analyze and decompose these inputs to develop an accurate understanding of their impact on the digital avionics system design requirements including the overall architecture and allocation of functions among the major system elements. Fortunately for the designer there are computer-based tools available to automate much of this task.

There are many standards, specifications, and characteristics that levy requirements on the design of the avionics system. One is of particular importance in the initial design phase, MIL-F-9490 Flight Control Systems—Design, Installation, and Test of Piloted Aircraft, General Specification for, since it imposes a host of tough requirements on the flight controls that must be completely understood by the designer from the start.

The Aircraft and Its Mission Drive the Avionics System Design

The starting point for designing a digital avionics system is a clear understanding of the mission requirements and the requirements levied by the host aircraft. If these requirements are not clearly understood, the avionics system will never achieve its intended performance. It is important to discuss how these top-level requirements are developed and to note the differences in development methods for civil and military avionics. For military aircraft the customer (Department, or Ministry, of Defense) prepares a statement of need and a top-level description of the possible mission(s) (e.g., air-to-air, reconnaissance, bombing, etc.), describes the gross characteristics of a hypothetical aircraft that could fly the mission, and then seeks vendors willing to build such an aircraft. The military customer may also describe the mission environment (e.g., day or night, all weather, and forward battle area) and define strategic and tactical philosophies and principles and rules of engagement.

Typical of the top-level requirements for a new aircraft are those shown in Table 1.1 for an aircraft that was to have been procured to replace the A-10 Thunderbolt. Of the requirements listed in Table 1.1, the avionics designer will pay particular attention to the takeoff and landing distance requirements for their potential impact on the flight controls. In addition, the requirements on turnaround time and the percentage of repairs that can be completed within a given time will directly affect the maintenance aspects of the design.

In 1987 the U.S. Air Force completed the definition of a generic avionics architecture for a hypothetical high-performance aircraft. This architecture is known as Pave Pillar, and some of its example performance requirements are given below:[1]

- Two-level maintenance
- Sortie rate: ≥4.5 per day
- Abort rate: ≤1 percent
- Combat turnaround time: ≤15 minutes (min)
- Nonmission capable (for avionics): ≤1.2 percent
- Mean time between critical failures:* 70 hour (h)
- Mean time to repair: (critical functions) 1.25 h (includes troubleshooting: 10 min)
- Fault detection: 99 percent of all possible faults
- Fault isolation: 98 percent of all possible faults

For civil transport aircraft there is a vastly different approach to defining the aircraft and its mission. The aircraft manufacturer makes a

---

*A critical failure is a failure of an essential system that would prevent the successful completion of the mission.

TABLE 1.1 Example Top-Level Requirements for the Successor to the A-10 Aircraft

	Minimum	Goal
Takeoff and landing distance	3000 feet (ft)	2000 ft
Turn rate (M = 0.7 to 0.9)	Trade-off 10–20°/s	
Mission radius at low altitude	250 nmi	
Mission loiter time	30 min	45 min
Turnaround time	25 min	15 min
Percentage of all repairs that can be completed within:		
2 h	60	65
4 h	75	80
8 h	85	95

very careful analysis of the potential customer's route structure, image, and operating philosophies to determine the customer's needs and postulates a future operating environment, especially air traffic management and relevant economic trends. The manufacturer then designs an aircraft that provides an optimum, balanced response to the integrated set of needs. Safety is always the highest-priority need and economical operation is a close second.

Figure 1.1 shows a portion of the overall system development cycle.[2] The present discussion deals with the first three blocks of the cycle. Later chapters will examine the remaining blocks.

Figure 1.2 shows the path from aircraft mission requirements to the avionics system requirements. The process begins by making a top-down analysis to the aircraft mission(s). If the aircraft is required to fly several types of missions, the most demanding mission should be selected as the baseline mission and all other possible missions treated as increments from the baseline. The baseline mission is divided into segments, and each segment is subjected to a detailed analysis to establish the requirements for the avionics system.

Standard mission segments include taxi, takeoff, climb, cruise, descent, landing, and rollout for both civil and military aircraft; however, rotorcraft missions do not generally include taxi and rollout. For both civil and military aircraft, diversion to an alternate airport is a likely mission segment and one that may contain some very tough requirements derived from operation in adverse weather. For military aircraft there are other possible mission phases, such as low-level penetration, air-to-air combat, air-to-ground attack, and reconnaissance. Each mission segment is examined at the lowest reasonable level to establish the avionics system performance requirements. Typical cruise phase require-

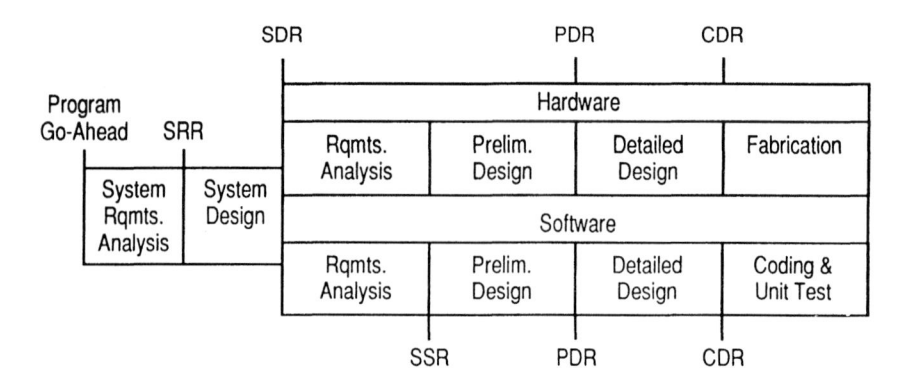

4   Chapter One

Figure 1.1 A portion of DOD-STD-2167A system development cycle.[2]

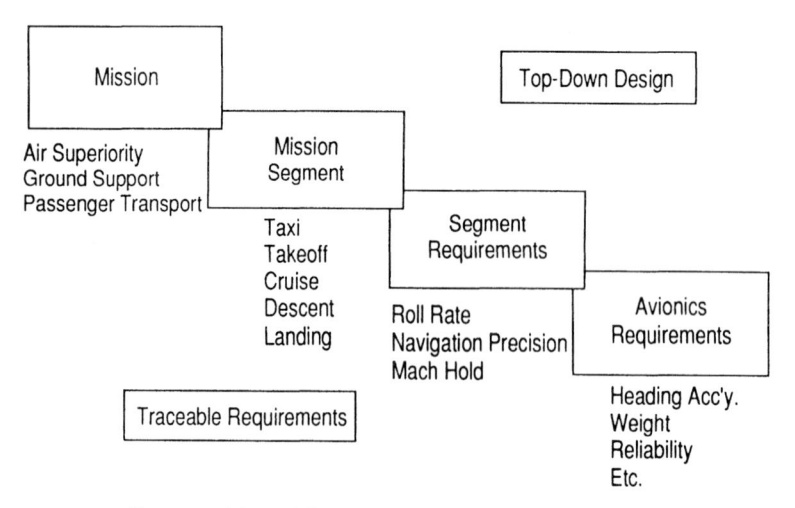

Key: Rqmts: Requirements
 SRR: System Requirements Review
 SDR: System Design Review
 SSR: Software Specification Review
 PDR: Preliminary Design Review
 CDR: Critical Design Review

Figure 1.2 Decomposition of the aircraft mission to establish the avionics system requirements.

ments may include attitude, altitude, and Mach hold; gust alleviation, crew (and passenger) comfort; and navigation. In a typical military mission such as an air-to-air mission, avionics requirements may include target range, range rate, azimuth, and so on. These requirements must be stated in quantitative, unambiguous terms.

The time the aircraft spends in a given mission segment must be accurately stated since it has a strong influence on the avionics system design. For example, the relatively brief time spent in the final approach and landing segment means only a short operating time for the automatic landing system. This brief time becomes a major factor in computing the probability of failure of the automatic landing system. Similarly, the amount of time spent in the air-to-air combat segment for high-performance aircraft is generally very short and will have a substantial impact on predicted probability of failure for equipment used only in that segment. Conversely, for example, if the aircraft spends an extended period in supersonic cruise, skin temperatures may be very high and subsequently may heat any nearby avionics.

Many of the requirements which influence the avionics system design are not directly related to the airborne performance of the aircraft. Overall aircraft performance requirements, such as maintenance hours per flight hour and mean time between failures (MTBF), must be examined and, if possible and appropriate, apportioned to the various systems on the aircraft. It is important to also note that some of the performance requirements may affect the design of the avionics ground support equipment, specifically requirements related to mean time to repair and ground turnaround time between flights.

While it may not be expressed as a requirement by the customer, the designer should be aware of the great importance placed by airlines on the mean time between unconfirmed removals (MTBUR) of avionics line replaceable units (LRUs). The MTBUR is not the same as MTBF since typically about half of the LRUs removed from aircraft in subsequent testing prove to be okay. Still it costs almost as much to remove and check a failure-free unit as it does to remove and repair a failed unit.

In order to achieve the maximum productivity and effectiveness and to reduce the aircraft life cycle cost, or cost of ownership, the customer may impose top-level requirements in terms of aircraft utilization rate. Examples include dispatch reliability for civil aircraft and mission capable rate for military aircraft.

It is essential that these mission and performance requirements be critically examined by the avionics system designer and other experts since a misinterpretation of these top-level requirements can lead to an error in the avionics system requirements and, consequently, to costly corrective redesign later.

Throughout the development and analysis of requirements, traceability is mandatory. Every requirement at every level of the design must be traceable to a requirement at the next higher level. Traceability becomes especially valuable when a particular requirement is proving difficult to meet. Knowing the genesis of a requirement may aid in its possible modification to make it more readily achievable.

As these requirements are developed, some of them may be identified as flight critical; that is, if the function being addressed by the requirement is not met, the aircraft will be lost. It is very important to note that if an avionics function is critical for one phase of flight, it must be treated as critical for all phases.

For military aircraft it is generally required that the contractor-developed requirements be formally reviewed and approved by the customer. This review and approval commonly occur at the System Requirements Review (see Fig. 1.2).

Mission Analysis Techniques

There are many mission analysis techniques that can be used to define, with varying degrees of rigor, the avionics system requirements that were discussed in the preceding section.[3] In a general sense these techniques implement the concept of functional decomposition.* A sample listing of these techniques, arranged in approximately chronological order according to design phase, includes mission profiles, mission scenarios, functional flow diagrams, decision/action diagrams, and function allocation trades. Because these techniques are neither mathematically precise nor rigorously defined, users may refer to them by a variety of names and may blend steps from several techniques, but the fundamentals are still the same. Some of these techniques have been automated to a certain extent with limited success.

Preparing a mission profile is generally the first step in defining the avionics requirements. A typical mission profile is a simple, annotated trace of the horizontal and vertical flight path of the aircraft. It is most appropriate for gross analysis and for laying the ground work for the detailed analyses in functional flow and allocation and decision/action diagrams.[3]

A mission scenario builds on the mission profile discussed in the previous paragraph. A scenario is generally a narrative description of subsystems and their proposed capabilities; sensors' (radar, infrared, electronic warfare, navigation, etc.) ranges and capabilities; armament; tactics; rules of engagement; weather; visibility; sea state; target/threat location, defenses, performance envelopes, tactics; and

---

*The term *functional decomposition* may be somewhat misleading and pretentious (like many technical terms). A more accurate and descriptive term is *function decomposition*, meaning the increasingly detailed analysis of a high-level, relatively imprecisely defined function to ultimately establish a series of smaller, quantitative functions.

geography of the anticipated combat area. A mission scenario should have enough detail to describe the activities performed in each mission phase, a preliminary definition of the accuracy required for each activity, a first-cut crew timeline, and interdependencies (sequence, coordination, and information transfer) and to determine the feasibility of proposed tactics. Developing a mission scenario is generally not a rigorous, structured process.[3]

For offensive and defensive avionics it is the target/threat that drives the requirements. The avionics system requirements must reflect the need for detection and recognition of the target/threat, armament range and capabilities (yours and theirs), and the host and threat aircraft maneuverability. The requirements must reflect an integrated perspective of the aircraft, armament, and avionics.

Functional flow diagrams, shown in Fig. 1.3, are a valuable and widely used mission analysis tool. Developing the diagrams is a much more structured and controlled process than that for mission profiles and mission scenarios and is designed to ensure traceability and consistency among evolving requirements at different levels. Functional flow diagrams are particularly well suited to allocating performance responsibilities among hardware, software, and/or the operator. Functional flow diagrams consist of blocks representing functions, each of

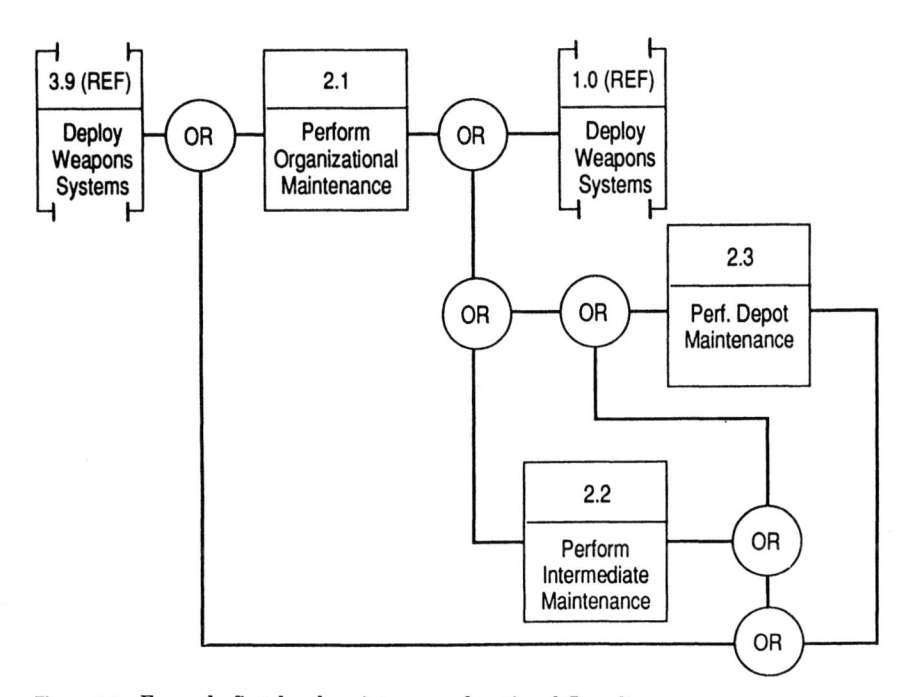

Figure 1.3 Example first-level maintenance functional flow diagram.

which is labeled with a noun and verb and has lines that depict the flow. Generally there are multiple, hierarchically related layers of diagrams. First-level diagrams are very similar across many programs. Second-level diagrams generally describe operational functions within mission phases. Third-level diagrams begin the function allocation among hardware, software, and the operator. In Fig. 1.3 the partially open rectangles, 3.9 and 1.0, refer to other functional flow diagrams. It is important to note that the diagrams do not show information flow or correlate information with the decisions implicit in the function boxes. Functional flow diagrams are best used in the early phases of a program.[3]

Decision/action diagrams are very similar to functional flow diagrams except that a binary decision block like those used in computer flow diagrams replaces the OR circle of the functional flow diagrams. Because of the similarity between them, the use of both functional flow diagrams and decision/action diagrams is not recommended. Only one of the two types of diagrams should be chosen. (It will come as no surprise to the astute reader, based on the clue given earlier, that the decision/action diagram is particularly well suited to software-intensive projects.)

As the next step in developing the avionics system requirements, the functions which have been identified in the previous steps must be assigned to one of the three basic components of any aircraft "system": hardware, software, and operator (crew) in a process known as function allocation trades. The assignment of functions is based on several factors: limitations of the operators, performance of the hardware and software, speed and accuracy requirements, and workload. There are three techniques for conducting the function allocation trade studies: trial and error (the way we have always done it), an evaluation matrix based on the Fitts list, and a design evaluation matrix.

The first of these techniques warrants no further discussion here except to note that modern, highly integrated avionics system probably cannot be very well designed doing it the way we have always done it.

The second technique is based on a seminal paper written by Paul Fitts over 40 years ago, one of the first in the field of human factors. Fitts sorted capabilities into two categories: those at which humans excel and those at which machines excel. Despite its age, the paper remains valid today with one possible exception: machines are now capable of learning from their experiences, albeit not with the speed or versatility of humans. This list, shown in Table 1.2, is a valuable guide in the top-level allocation of functions. The list can be used to develop a function allocation screening worksheet, an example of which is shown in Fig. 1.4, to permit a quantitative basis for function allocation. In Fig. 1.4, each identified function is listed in the left col-

TABLE 1.2 Fitts List Comparing Human versus Machine Capabilities[4]

People excel in	Machines excel in
Detection of certain forms of very low energy levels	Monitoring (both people and other machines)
Sensitivity to an extremely wide range of stimuli	Performing routine, repetitive, or very precise operations
Perceiving patterns and making generalizations about them	Responding very quickly to control signals
Ability to store large amounts of information for long periods and recalling relevant facts at appropriate times	Storing and recalling large amounts of information in short time periods
Ability to exercise judgment where events cannot be completely defined	Performing complex and rapid computation with high accuracy
Improvising and adopting flexible procedures	Sensitivity to stimuli beyond the range of human sensitivity (infrared, radio waves, etc.)
Ability to react to unexpected, low-probability events	Doing many different things at one time
Applying originality in solving problems (i.e., alternative solutions)	Exerting large amounts of force smoothly and precisely
Ability to profit from experience and alter course of action	Insensitivity to extraneous factors
Ability to perform fine manipulation, especially where misalignment appears unexpectedly	Ability to repeat operations very rapidly, continuously, and precisely the same way over a long period
Ability to continue to perform when overloaded	Operating in environments which are hostile to humans or beyond their tolerances
Ability to reason inductively	Deductive processes

umn and assigned a weighting factor, based on the judgment of the designer, corresponding to its importance. Across the top of the form each capability is listed as either an operator or an equipment capability and is also assigned a weighting factor, again based on the judgment of the designer, according to its importance. A score, the product of the importance and capability factors, is calculated for each element in the matrix and a total score for each function (row) is determined for operator and machine capabilities. If either the operator or the machine has a clearly higher score, the function is assigned to that capability. Where there is little difference in the total scores between the two capabilities, the function is assigned on the basis of other factors such as workload of the candidate element when the function is being performed.

KEY:

5/25 ← weighted score
↑
rated score

SCALE 1-5 5 BEST

HYPOTHETICAL TRACKING FUNCTIONS	INHERENT OPERATOR CAPABILITIES					INHERENT EQUIPMENT CAPABILITIES			TOTAL SCORE		PROPOSED ALLOCATION			
	DETECTING SIGNALS IN THE PRESENCE OF HIGH NOISE ENVIRONMENT (X5)	RECOGNIZING OBJECTS UNDER VARYING CONDITIONS OF PERCEPTION (X4)	HANDLING UNEXPECTED OCCURRENCES OR LOW-PROBABILITY EVENTS (X4)	REASONING INDUCTIVELY (X1)	PROFITING FROM EXPERIENCE (X2)	RESPONDING QUICKLY TO SIGNALS (X3)	PERFORMING PRECISE ROUTINE REPETITIVE OPERATIONS (X2)	COMPUTING AND HANDLING LARGE AMOUNTS OF STORED INFORMATION QUICKLY AND ACCURATELY (X4)	OPERATOR	MACHINE	OPERATOR	BOTH	MACHINE EQUIPMENT	MACHINE SOFTWARE
1. DETERMINE IF TARGET TRACKS IN SYSTEM	5/25	2/8	3/12	3	3/6	3/9	4/8	1/4	81	41	X			
2. ACTUATE SEQUENCE	1/5	1/4	1/4	1	1/2	1/3	1/2	1/4	20	24		X		
3. PUT NEXT TARGET IN TRACK LIST UNDER CLOSE CONTROL	1/5	1/4	1/4	1	1/2	3/9	5/10	1/4	21	43			X	X
4. ADVANCE HOOK ON DISPLAY TO TRACK COORDINATES	1/5	1/4	1/4	1	2/4	3/9	5/10	1/4	70	39	X			
5. DETERMINE IF HOOK LINES UP WITH PRESENT TARGET POSITION	4/20	2/8	2/8	3	2/4	3/9	4/8	1/4	73	40	X			
ETC....														

Figure 1.4 Example evaluation matrix based on the Fitts list.[3]

The third function allocation technique, the design evaluation matrix, is very similar to the function allocation worksheet in the previous paragraph. Figure 1.5 is an example design evaluation matrix in which there are three trade options under study and ten evaluation characteristics with weighting factors, assigned by the designer, ranging from 1 to 10. Each cell is assigned an index factor ranging from 1 for unfavorable to 5 for favorable (when compared to the other options). The score for each cell is the product of the weighting and cell index factors. The total score for each option is the sum of the cell scores for all ten characteristics. In this example the two crew members option with a total score of 157 would be selected.[3]

There are some mission analysis techniques that have been automated and used successfully in the design of actual digital avionics systems. One example is Semi-Automated Functional Requirements Analysis (SAFRA) developed by the British Aerospace Co. Military Aircraft Division to design the avionics system on the Experimental Aircraft Program (EAP), the forerunner of the Eurofighter Aircraft (EFA). SAFRA "encompasses a number of methods and tools which support the various stages of the system-software life cycle." The underlying fundamental principle of SAFRA is "progressive decomposition of high level requirements in a logical and consistent manner until a level is reached where the requirements are expressed in sufficiently precise detail to allow hardware and software design to commence."[5,6] The heart of SAFRA is the controlled requirements expression (CORE) method for producing unambiguous, consistent, and complete system and software requirements.

SAFRA, shown in Fig. 1.6, is divided into three stages, each of which has inputs, activities, and outputs. Stage 1 operates on inputs of the air vehicle specification and general tactical principles and philosophies to produce a functional statement without reference to mechanization or location of the function. The functional statement defines information flows to and from the avionics system and partitions functions at the flight control, utilities, and avionics system level. Stage 2 uses this functional statement and a safety analysis to (1) establish subsystem functional requirements, (2) establish data attributes, transfer rates, and word formats, (3) determine overall sequence of operation and iteration rate, and (4) partition within a system (e.g., avionics, to communications, navigation, displays, etc.). Stage 3 uses subsystem functional requirements and interface documents from stage 2 to produce detailed hardware and software requirements. Stage 3 also defines software schedules and overall code structure and details of the process algorithms.

The key to the success of SAFRA is the rigorous, formal configuration and change control that is intrinsic to a computer-based method-

WEIGHTING FACTOR	10	5	5	4	4	4	3	3	2	1	
EVALUATION CHARACTERISTICS	Cost	Autonomy	Response time	Multimission flexibility & growth	Reliability	Safety	Rescue capability & EVA	Replanning	Operating weight	Other	SCORE
One crewmember	5 / 50	2 / 10	5 / 25	3 / 12	2 / 8	2 / 8	1 / 3	1 / 3	5 / 10	2 / 2	131
Two crewmembers	4 / 40	4 / 20	4 / 20	4 / 16	3 / 12	3 / 12	4 / 12	4 / 12	4 / 8	5 / 5	(157)
Four crewmembers	1 / 10	4 / 20	3 / 15	5 / 20	3 / 12	3 / 12	4 / 12	4 / 12	4 / 8	1 / 1	122

OUTCOME: two crew member option by 20%

CELL INDEX FACTOR KEY
1 = unfavorable
2 = slightly unfavorable
3 = neutral
4 = favorable
5 = very favorable

- Evaluation chracteristics are analyzed with respect to crew considerations only

WEIGHTING FACTOR KEY
1 = low weight
10 = high weight

TRADE OPTIONS

Figure 1.5 Example design evaluation matrix.³

Establishing the Avionics System Requirements 13

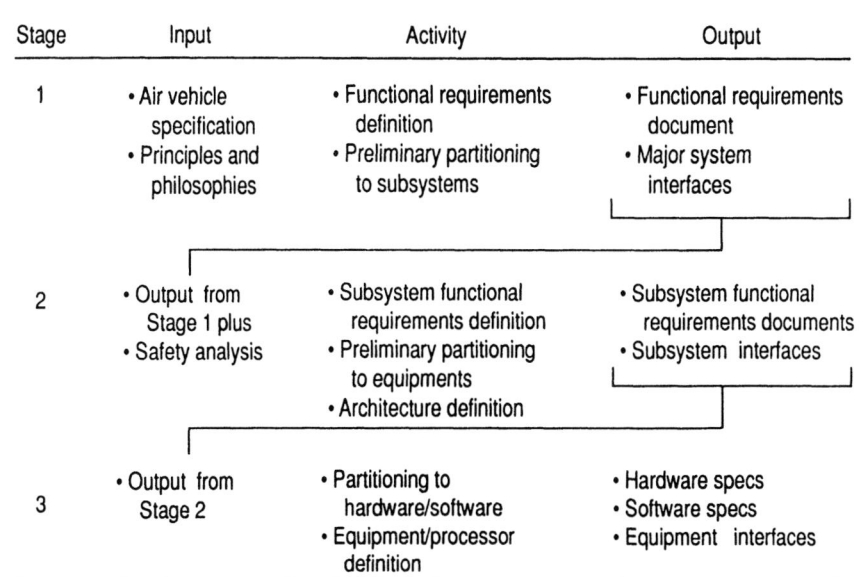

Stage	Input	Activity	Output
1	• Air vehicle specification • Principles and philosophies	• Functional requirements definition • Preliminary partitioning to subsystems	• Functional requirements document • Major system interfaces
2	• Output from Stage 1 plus • Safety analysis	• Subsystem functional requirements definition • Preliminary partitioning to equipments • Architecture definition	• Subsystem functional requirements documents • Subsystem interfaces
3	• Output from Stage 2	• Partitioning to hardware/software • Equipment/processor definition	• Hardware specs • Software specs • Equipment interfaces

Figure 1.6 Semi-automated functional requirements analysis (SAFRA) methodology.[4]

ology. Any successful mission analysis methodology must give first priority to configuration and change control and must be user friendly. User acceptance of these computer-based methodologies is essential—if the designer refuses to provide correct and current inputs and to recognize the output as authoritative and correct, the methodology is doomed to failure.

In Chaps. 9 and 10 additional computer-aided design tools will be discussed. These highly automated tools, which generally require as inputs the type of documentation generated in the first stage of SAFRA, use rigorous techniques to ensure consistency and compatibility among the system elements as the design evolves, and they impose tight configuration control. The more capable of these tools can also generate project documentation and software in several high-level languages.

A Look at the Requirements of MIL-F-9490

In many aircraft the most important avionics function is flight controls, frequently in a flight critical manner; that is, if the avionics fail, the aircraft is lost. Because of this critical role for avionics, it is essential to take an early look at military specification MIL-F-9490: Flight Control Systems—Design, Installation and Test of Piloted Aircraft, General Specification for, the benchmark for the application of digital avionics to flight control systems for U.S. Air Force aircraft and rotorcraft.[7] Although 9490 is not applicable to civil aircraft, many of its principles and philosophies could be generically applied to civil aircraft flight control systems. As the title suggests, 9490 is a general specification and applies to all mechanical, analog, and digital flight control systems. This discussion will only focus on selected general requirements and those requirements unique to digital flight control systems. Since flight control is the most demanding application of digital avionics, 9490 levies some very tough requirements.*

MIL-F-9490 establishes several definitions generally not found in specifications dealing with avionics. There are five operational states for the flight control system:

Operational state I: Normal operation.

Operational state II: Restricted operation. Degradation or failure of noncritical portion of the overall flight system. Moderate increase in crew workload or reduced mission effectiveness, but the mission can be accomplished.

Operational state III: Minimum safe operation. Degraded flight control system performance, safety, or reliability. Excessive crew workload and/or sharply reduced mission effectiveness. Can return safely and land at original destination or an alternate one.

Operation state IV: Controllable to an immediate emergency landing.

Operation state V: Controllable to an evacuable flight condition.

A further set of definitions for criticality classification is established by 9490. These definitions do not match definitions for similar terms found in Federal Aviation Regulation (FAR) 25.1309 (see Chap. 9), so it is important to specify which usage is intended, especially in the case of the term *essential*.

Essential: A function is essential if its loss degrades the flight control system performance beyond Operational State III.

Flight phase essential: The same as essential above except it applies only during specific flight phases.

Noncritical: Loss of function does not affect flight safety or reduce control capability beyond that required for Operational State III.

---

*MIL-F-9490 has been superseded by MIL-F-87242 Flight Control System, General Specification for, 31 March 1986, which has limited distribution and, thus, is not publicly available. One can assume, however, that MIL-F-87242 will incorporate many of the generic requirements found in MIL-F-9490. Even though 9490 has been superseded, the latest aircraft, such as the C-17 continue to use it.

The term *extremely remote* is used to describe the probability of failures in the flight control system in a manner similar to FAR 25.1309. The precise value of the term depends on the class of aircraft, as shown below:

Class of aircraft	Probability of failure (per flight hour)
Heavy bomber, transports, cargo, and tankers	$\leq 5 \times 10^{-7}$
Rotorcraft	$\leq 25 \times 10^{-7}$
All other aircraft	$\leq 100 \times 10^{-7}$

In some places the specification is a paragon of lucidness. For example, paragraph 3.1.3 states, "Flight control systems shall be as simple, direct, and foolproof as possible, consistent with overall system requirements" and paragraph 3.1.3.4 states, "Systems shall be arranged to satisfy the reliability, invulnerability, failure immunity, and other general requirements of this specification." These statements are probably the most sound design advice found in any military specification, and they can be applied to almost any hardware.

Since many of the requirements in 9490 are also excellent design goals for any avionics system, they have been incorporated, generally in modified form, in many locations throughout this book. However, these same requirements also will be examined here to present a coherent synopsis of 9490.

Survivability and invulnerability are prominent among the "ilities" discussed in 9490. The specification for survivability simply requires that the flight control system shall be capable of maintaining at least Operational State IV performance in the absence of engine thrust.

Invulnerability specifications place larger demands on the system design. According to 9490, varying degrees of flight control system performance degradation are permissible for a host of external hazards. These hazards include natural and induced environments, lightning and atmospheric static electricity, failure of other onboard systems, maintenance or flight crew inaction or error, or enemy action.

In the case of induced environments, MIL-E-6051 Electromagnetic Compatibility Requirements, Systems, and MIL-STD-461 Electromagnetic Interference Characteristics Requirements for Equipment provide detailed requirements for the flight control system in terms of generating and tolerating electromagnetic interference (EMI). Some of the requirements in MIL-STD-461 are examined in Chap. 9. In any induced environment, the flight control system must be capable of at least maintaining Operational State II capability. Temporary (no exact time is stated) reduction to Operational State III is permitted.

In the presence of lightning or atmospheric electricity, the flight control system must be capable of maintaining Operational State II capability with temporary reduction to Operational State III in the case of a direct lightning strike.

The "failure of other onboard systems" hazard levies more requirements. The flight control system must maintain at least Operational State III capability after failure of the critical engine in a two-engined aircraft or failure of the two most critical engines in an aircraft having three or more engines. This Operational State III is in addition to the Operational State IV requirement in the event of the loss of all engine thrust, as discussed under vulnerability. Other strict requirements include (1) for heavy bombers, transport, cargo, and tanker aircraft, the probability of experiencing loss of flight control system performance below Operational State IV shall be extremely remote ($\leq 5 \times 10^{-7}$/flight hour) for engine or other rotor burst and (2) for all other aircraft this same probability shall apply to Operational State V.

Invulnerability to a hazard caused by enemy action requires that the flight control system must maintain at least Operational State II after at least one encounter from a threat as defined by the procuring activity. Primary structural components are one means that can be used to provide protection necessary to meet this requirement.

In addition to the requirements for tolerating the various hazards described above, 9490 places other noteworthy demands on the flight control system. For example, multiplexed data buses must conform to MIL-STD-1553 (see Chap. 2). Sound design practice calls for redundant data buses that are electrically and spatially isolated to the maximum extent possible. Only one bus shall be vulnerable to loss by an event that has a probability worse than extremely remote.

Most designers prefer to use new technologies in the design of avionics systems. However, in many cases 9490 requires that specific approval be obtained before including new technologies in essential elements of the system. Some of the technologies subject to that provision include optical transmission media for data and nonhydraulic control surface actuators (electromechanical, motor-pump servoactuators, and pneumatic).

Wherever critical or flight-phase critical elements are interfaced with noncritical elements, provisions shall be made for isolation and separation so that the probability of common mode failures between different classes of criticality elements is extremely remote.

MIL-F-9490 recognizes the need for growth provisions. Such provisions ensure that the flight control system can take full advantage of the intrinsic flexibility and expandability of digital systems. When the aircraft is accepted by the procuring activity, (1) the total time used for all computations by the flight control system in the worst case shall not exceed 75 percent of the total time available, and (2) no more than 80 percent of the available memory shall have been committed.* The computation and sampling rates must be such that they do not introduce unacceptable phase shift, round-off error, nonlinear characteristics, or aliasing into the system response.

For high-maneuverability aircraft, the warm-up time for the flight control system shall be 90 seconds (s) or less. For all other aircraft, it shall be 3 min or less.

Because of the criticality of the flight control system, a system development plan must be submitted to the procuring activity for review and approval. The overall plan must contain the following: (1) a detailed milestone chart, (2) a synthesis and analysis plan, (3) a verification plan, (4) flight safety, reliability, maintainability, and vulnerability analyses plans, (5) a functional mockup test plan, (6) a ground test plan including test procedures, and (7) a flight test plan including test procedures. A flight control system specification must also be prepared.

Concluding Remarks

Understanding and correctly interpreting the top-level performance requirements is the first step in designing and building an avionics system that will be a star performer. Nonoperational requirements, such as MTBF, are as important as the more familiar and visible operational requirements. A top-down approach, commonly known as functional decomposition, to developing the avionics system requirements is strongly recommended. There are many functional decomposition methodologies available, and the designer should choose one that best suits his or her skills and the design needs at that time. Developing the avionics requirements is a balancing act between the pull of performance demands and the restraints of regulations and specifications.

Three topics that often are not given adequate attention in the early stages of design are maintenance, test, and interfaces with the host aircraft. Neglect of these topics in the early design phases almost ensures that they will cause problems later in the development and deployment of the system. In order to give them the attention they deserve, they are discussed in detail in Chaps. 6 and 8.

---

*Because of the relatively low cost of microprocessors and memory chips and the inevitable growth in avionics system functions as the system design evolves, modern practice calls for only 50 percent or less use of the total computational and memory capability at the time of the Preliminary Design Review.

References

1. AFWAL-TR-87-1114 "Architecture Specification for PAVE PILLAR Avionics," January 1987.
2. DOD-STD-2167A Defense System Software Development, 29 February 1988.
3. DOD-HDBK-763 Human Engineering Procedures Guide, 27 February 1987.
4. Fitts, Paul M. (ed.), "Human Engineering for Effective Air-Navigation and Traffic-Control System," National Research Council, Washington, 1951, pp. 5–11.
5. Malley, H. M., Jewell, N. T., and Smith, R. A. C., "A Structured Approach to Weapon System Design," *AGARD Conference Proceedings no. 417,* April 1987.
6. Rowley, J. D.: "A Structured Approach to Weapon System Design" (updated), *AGARD Lecture Series no. 164,* May 1989.
7. MIL-F-9490: Flight Control Systems—Design, Installation and Test of Piloted Aircraft, General Specification for, 6 June 1975.

Bibliography

Frenock, T. J., "Advanced Integrated Flight Control System," Boeing Military Airplane Co. NASA Contract NAS2-11866 (1985).

Chapter

2

Avionics Systems Essentials I: Data Buses

Just as our body needs a central nervous system, so do aircraft. In the case of modern aircraft, a data bus is the central nervous system, but, unlike our nervous systems, signals travel much faster and redundancy can negate damage effects.

Multiplexing, the transmission and reception of multiple signals over a common path, is one of the cornerstones in an integrated digital avionics system. It allows the sharing of data and computation results, thereby ensuring that all connected subsystems are using a consistent (and presumably correct) database while reducing the weight of the wiring. Multiplexing also enables highly flexible system designs that can grow and change as the mission of the aircraft changes and new avionics capabilities are developed and installed.

Since sharing of data is implicit in multiplexing, it is essential that data bus hardware and signals adhere to a common, versatile standard. Nearly two decades of experience with data buses and a generous amount of time and help by volunteer committees are manifest in the most commonly used data bus standards, MIL-STD-1553 and Aeronautical Radio, Inc., (ARINC) Specification 429.

Before examining these and other data bus standards, it is important to note that there are some instances where, based on signal integrity and/or system complexity, signals should not be multiplexed. Examples include discrete signals from gear position, stores armed or fired, or cargo door ajar. Also, high-priority or extremely urgent signals such as engine fire, nuclear event detector, or ground proximity warning, traditionally have been hardwired but may be transmitted on high-integrity buses between systems and line replaceable units (LRUs). Simple warnings, such as lights, should continue to be hardwired.

Data Buses Using MIL-STD-1553

MIL-STD-1553 Digital Time Division Command/Response Multiplex Data Bus establishes the requirements for digital data buses on military aircraft. The Foreword of the standard states that it includes "techniques which will be utilized in systems integration of aircraft systems." The use of 1553 is common on all new military aircraft, and it is frequently used in major avionics retrofit programs. Even though it is designated as an aircraft internal standard, it is spreading in use to include the aircraft weapons systems and even some ground-based equipment. There is general agreement that the use of MIL-STD-1553 will continue to grow since it offers an opportunity to make giant strides toward integrated systems.[1]

The standard is divided into three parts:

1. Types of terminals—bus controller, bus monitor, and remote terminal
2. Bus protocol, including message formats and word structure
3. Hardware performance specifications, such as characteristic impedance, operating frequency, signal droop, and correction requirements

Figure 2.1 is a hypothetical design of a simple MIL-STD-1553 data bus which operates at 1 megabit per second (Mbit/s) in a half-duplex mode using Manchester II data encoding.

Hardware is the tangible portion of a 1553 bus. As stated above, the major hardware elements are the bus controller, remote terminal, and an optional bus monitor.

The bus controller is in charge of all data flow on the bus and initiates all information transfers. The bus controller also monitors the status of the system. However, this latter role is not to be confused with the bus monitor described in the next paragraph. Although the standard has provisions for the bus control capability to be transferred among terminals, Air Force Notice 1 prohibits using the bus control transfer capability. Bus controllers are generally separate LRUs but they can be part of one of the other LRUs on the bus.

A bus monitor, according to the standard, receives and stores selected bus traffic. A bus monitor will not respond to any traffic received unless the traffic is specifically addressed to it. Bus monitors are generally used to receive and extract data for off-line purposes such as flight test, maintenance, or mission analysis.

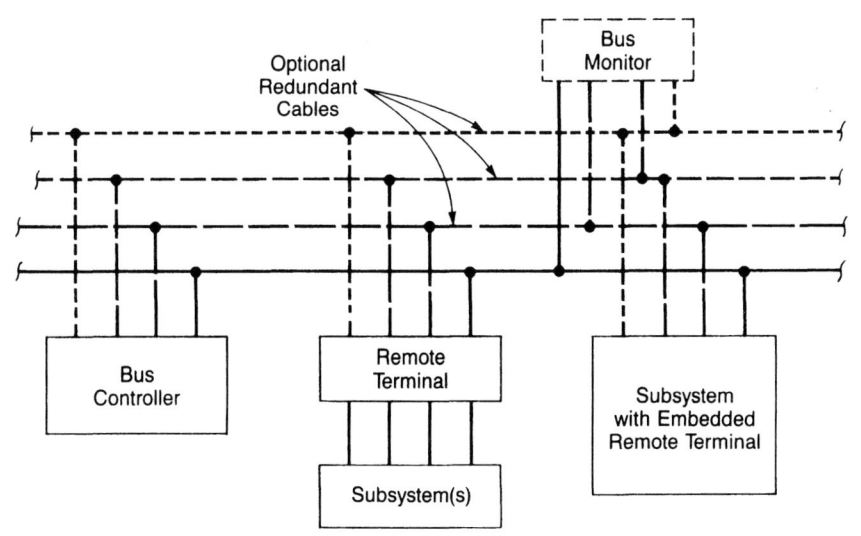

Figure 2.1 Typical MIL-STD-1553 bus structure.[1]

Remote terminals (RTs) are the largest fraction of units in a 1553 bus system. Because of the 5-bit RT address field, there can be only up to 31 RTs on a given bus. RTs respond only to valid commands specifically addressed to them or to valid broadcast (all RTs simultaneously addressed) commands. The address of a RT is established by wiring in an external connector. This approach allows any LRU of a given type to be used in any location without making internal changes to the LRU. No single failure in a RT shall cause it to validate a false address. As shown in Fig. 2.1, the RT can be separate from the subsystem(s) it serves, or it can be embedded in it or them.

Messages, of course, are the essence of any communications scheme. MIL-STD-1553 protocols allow 10 message formats or "information transfer formats." Each message contains at least two words, each of which has 16 bits plus synchronization and parity covering 20 bit times or 20 microseconds (μs). However, before discussing the message formats, the three types of allowable words that can comprise the messages must be reviewed.

All words are constructed using Manchester coding and have odd parity. A logical 1 begins positive and transitions to negative at mid-bit and a logical 0 begins negative and transitions to positive at mid-bit. Manchester coding was chosen since it is compatible with transformer coupling and is self-clocking. For all 1553 words, an invalid Manchester of 2 bits, covering the first 3 bit times, serves as a synchronization code as shown in Fig. 2.2. Unused bits are set to logical zero.

A command word is always the first word in a message and is transmitted only by the bus controller. The format of a command word is shown in Fig. 2-2a. Following the synchronization code, which is a

22 Chapter Two

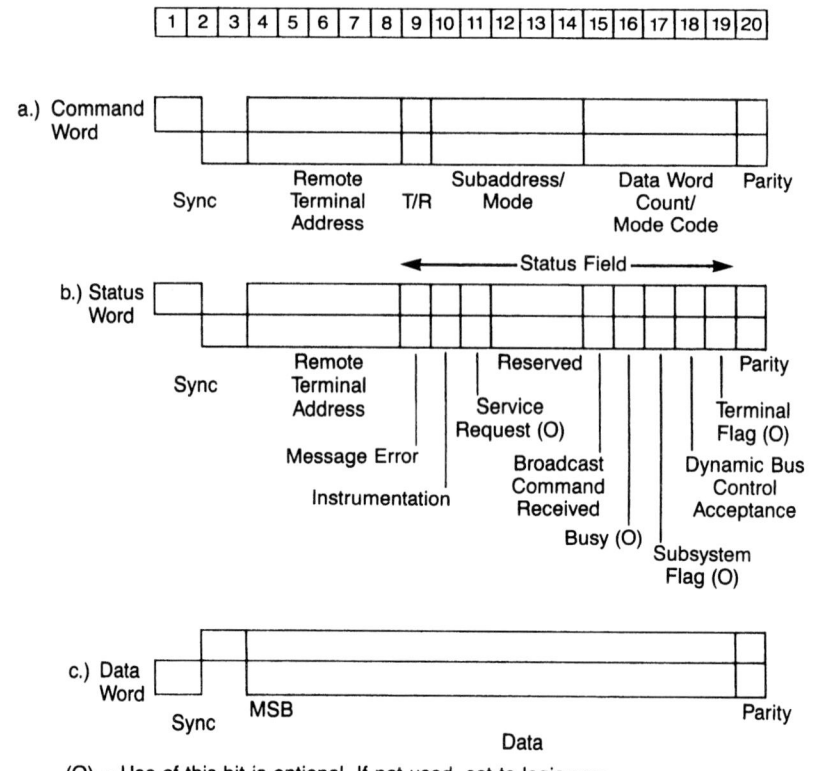

Figure 2.2 MIL-STD-1553 word formats.[1]

+1½ bit times followed by a −1½ bit times, there are 5 address bits. Every RT must have a unique address. Decimal address 31 (11111) is reserved for the broadcast mode, and every RT must recognize 11111 as a legal address (if the broadcast mode is used) in addition to its unique address. The transmit/receive bit is set to logical 0 if the RT is to receive and to logical 1 if the RT is to transmit.

The next 5 bits (10–14) are used to designate a subaddress to the RT or use of mode commands to the equipment on the bus; 00000 or 11111 is used to indicate mode codes will follow, which leaves a balance of 30 valid subaddresses. All RTs must be capable of responding to mode codes (00010 through 01000. If the instrumentation bit in the status field is used, the available subaddresses are reduced to 15.

If bits 10 through 14 are subaddresses, bits 15 through 19 are the data word count. Up to 32 data words can be specified to be transmitted in any one message. If bits 10 through 14 are either 00000 or 11111, bits 15 through 19 are mode codes. Mode codes are used only to communicate with the bus hardware and to manage information flow, not to transfer data. Table 2.1 gives the assigned mode codes. MIL-STD-1553 provides a thorough discussion of the meaning of each of the mode codes. Finally, bit 20 is the parity bit. MIL-STD-1553 requires odd parity for all three types of words.

A status word is always the first word in a response by a remote terminal. Figure 2.2b shows the format for a status word and its four basic parts: synchronization, remote terminal address, status field, and parity bit.

Bit positions 1 through 3 are the synchronization code that is identical to that of a command word (positive the first 1½ bit times and then negative the second 1½ bit times. Bits 4 through 8 are the address of the terminal transmitting the status word.

Bits 9 through 19 are the RT status field. All bits in the field are set to logical zero unless the named condition exists. The meaning of most bits is obvious; however, several require additional explanation. Bit 9 shall be set to logical 1 if one or more words in the preceding transmission from the bus controller was invalid. The instrumentation bit, in position 10, always shall be set to a logical 0 to allow distinction of

TABLE 2.1 Assigned Mode Codes.[1]

T/R bit	Mode code	Function	Associated data word?	Broadcast command allowed?
1	00000	Dynamic bus control	No	No
1	00001	Synchronize	No	Yes
1	00010	Transmit status word	No	No
1	00011	Initiate self-test	No	Yes
1	00100	Transmitter shutdown	No	Yes
1	00101	Override transmitter shutdown	No	Yes
1	00110	Inhibit terminal flag bit	No	Yes
1	00111	Override inhibit terminal flag bit	No	Yes
1	01000	Reset remote terminal	No	Yes
1	01001 ↓	Reserved	No	TBD
1	01111	Reserved	No	TBD
1	10000	Transmit vector word	Yes	No
1	10001	Synchronize	Yes	Yes
1	10010	Transmit last command	Yes	No
1	10011	Transmit bit word	Yes	No
0	10100	Selected transmitter shutdown	Yes	Yes
0	10101	Override selected transmitter shutdown	Yes	Yes
1 or 0	10110 ↓	Reserved	Yes	TBD
1 or 0	11111	Reserved	Yes	TBD

a status word from a command word. (In a command word bit 10 is always set to a logical 1 if the instrumentation bit is being used.) Using bit 10 as an instrumentation bit reduces the possible number of subsystem addresses to 15, as described in the command word discussion. The subsystem flag in bit position 17 is used to indicate a subsystem fault condition. Bit position 19, if used, indicates a fault condition within the RT (vis-à-vis bit 17 indicating a fault in a subsystem). Bit 20 is the parity bit.

A data word is the simplest, most straightforward of the three word types. Data words always follow command, status, or other data words; they are never transmitted first in a message. Like command words and status words, data words are 20 bit times long; the first 3 bit times are the synchronization code, and the last bit time is the parity bit (see Fig. 2.2c). However, the synchronization code for a data word is the reverse of that for command and status words: The first 1½ bit times is negative and the last 1½ bit times is positive. The remaining 16 bit times, 4 through 19, are for the binary coded data value. The most significant bit of the value is transmitted first. All unused bits are set to logical zero. The designer, if possible, should always strive to efficiently use all 16 bits by bit-packing multiple parameters and words.

As stated earlier, there are 10 allowable message formats, shown in Table 2.2. All messages must conform to one of the formats. In Table 2.2 the allowable response time is 4 to 12 μs and the intermessage gap is at least 4 μs. The minimum no response time out (which will set an LRU failure flag) is 14 μs. A message may contain a maximum of 32 data words.

All of the formats are executed under the direct control of the bus controller, and the first six in Table 2.2 require a specific, unique response from the RT being addressed. The last four formats are broadcast formats, which allow a RT to transmit a message to all addresses on the bus without the receiving terminals acknowledging receipt. Although this sounds like a very attractive approach, MIL-STD-1553 strongly advises against the use of this capability since errors and failure to receive the message would not be detected. The broadcast mode is not expressly prohibited, but any use of it must be clearly and thoroughly justified.

The hardware characteristics are the most straightforward parts of the standard. They are written in familiar specification language and are much easier to interpret than some other portions of the standard. Detailed characteristics are provided in paragraph 4.5 of the standard and are summarized below.

To begin with, the cable shall be two strands, twisted, shielded, and jacketed. There shall be four twists per foot, and the shielding shall

TABLE 2.2 MIL-STD-1553 Information Transfer Formats.[1]

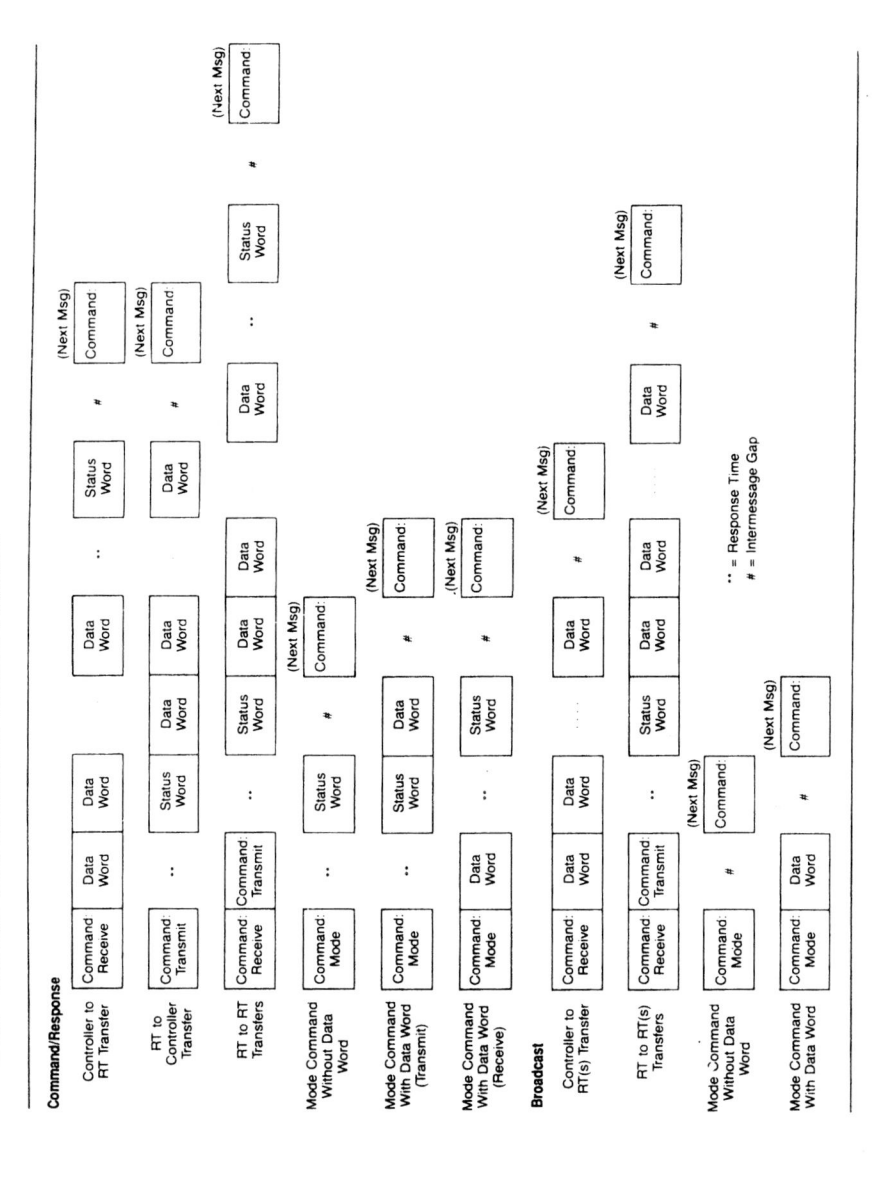

Command/Response

Controller to RT Transfer: Command Receive | Data Word | Data Word | ... | Data Word | Status Word | (Next Msg) Command | #

RT to Controller Transfer: Command Transmit | .. | Status Word | Data Word | Data Word | ... | Data Word | # | (Next Msg) Command

RT to RT Transfers: Command Receive | Command Transmit | .. | Status Word | Data Word | Data Word | ... | Data Word | .. | Status Word | # | (Next Msg) Command

Mode Command Without Data Word: Command Mode | .. | Status Word | # | (Next Msg) Command

Mode Command With Data Word (Transmit): Command Mode | .. | Status Word | Data Word | # | (Next Msg) Command

Mode Command With Data Word (Receive): Command Mode | Data Word | .. | Status Word | # | (Next Msg) Command

Broadcast

Controller to RT(s) Transfer: Command Receive | Data Word | Data Word | ... | Data Word | (Next Msg) Command | #

RT to RT(s) Transfers: Command Receive | Command Transmit | .. | Status Word | Data Word | Data Word | ... | Data Word | # | (Next Msg) Command

Mode Command Without Data Word: Command Mode | # | (Next Msg) Command

Mode Command With Data Word: Command Mode | Data Word | #

.. = Response Time
# = Intermessage Gap

cover 90 percent of the cable surface. The characteristic impedance at 1.0 megahertz (MHz) shall be between 70 and 85 ohms (Ω). Each end of the cable must be terminated in a resistor equal to the characteristic impedance ±2 percent. Wire-to-wire capacitance shall be ≤30 picofarads per foot (pF/ft) and cable loss shall be ≤1.5 decibels (dB) per 100 ft at 1.0 MHz. There is no limit on cable length.

Two ways to allow coupling are described in the standard. The first and most obvious is direct wire-to-wire connection, known as a direct coupled stub. Although the characteristics of a direct coupled stub, including a maximum allowed length of 1 ft, are described in the standard, Air Force Notice 1 to the standard suggests that direct coupled stubs should be avoided if at all possible because of the risk to the entire bus if a direct coupled stub were to develop a short circuit. Thus, in effect, the designer is left with only the second coupling method—transformer coupled stubs.

Figure 2.3 is a standard transformer coupled stub. There are still wire-to-wire connections; however, they now pass through isolation resistors to a coupling transformer. The resistance of each isolation resistor is $0.75Z_o$ ±2 percent, and the coupling transformer has a 1.4:1

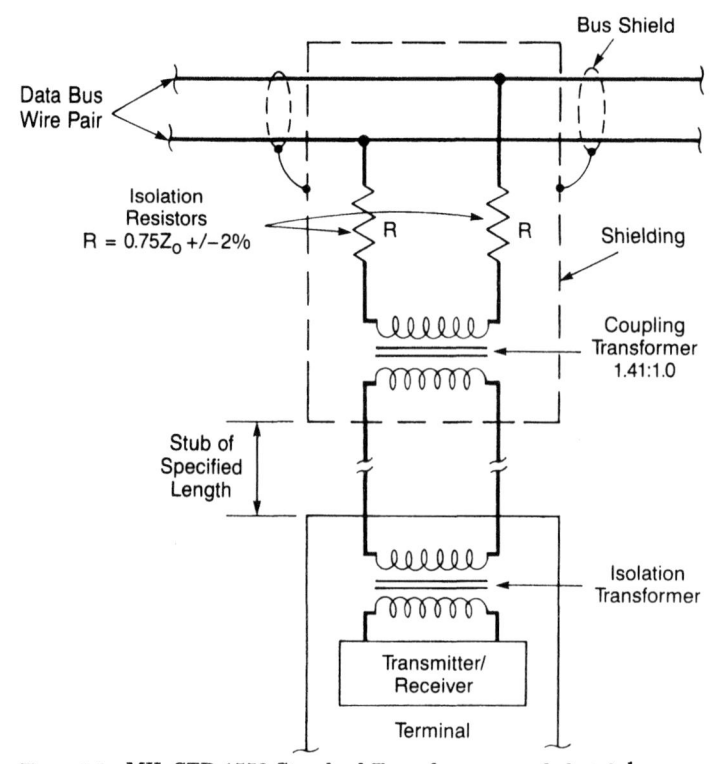

Figure 2.3 MIL-STD-1553 Standard Transformer coupled stub.[1]

turn ratio with the high side connected to the isolation resistors. Transformer coupled stubs can be any length, but the designer should strive to keep them under 20 ft if possible. Common mode rejection shall be ≥45 dB.

MIL-STD-1553 also includes detailed performance specifications for the terminals. The essence of this part of the specification is (1) the output voltage for transformer coupled terminals shall be 18 to 27 volts (V) peak-to-peak with less than 14 millivolts (mV) root mean square (rms) noise and (2) transformer coupled terminals shall respond to an input of 0.84 to 14.0 V peak-to-peak and have a minimum input impedance of 1000 Ω for input frequencies from 75 kilohertz (kHz) to 1 MHz.

DOD-STD-1773 Is the Fiber Optic Companion to MIL-STD-1553

DOD-STD-1773 is the optical equivalent of MIL-STD-1553 and was developed to realize the optical technology advantages, when used in a digital avionics system, of reduced weight, larger bandwidth, and immunity to electromagnetic interference (EMI).[2] DOD-STD-1773 has the same word structure and length and bus protocol as MIL-STD-1553. It is implemented as a 1-Mbit/s bus, but the standard acknowledges the feasibility and merits of a much higher bandwidth bus. While 1553 establishes in complete detail the signal characteristics, 1773 only refers the reader to the system specification in which 1773 is being applied. The same is also true for the cable and connector characteristics; to wit, "The optical power levels, optical wavelength and the means of distributing optical power in any specific implementation must be contained in a specification which references this standard." This omission of definition of key features of 1773 buses reflects the rapid improvements in optical technology. Including these requirements in the specification for the system in which 1773 is being used ensures that the best available technology is used when the system is built.

Since 1773 deals with optical rather than electrical signals, changes have to be made to the Manchester II coding. A logical 1 begins with optical energy present, which is then turned off at mid-bit. Similarly, a logical 0 begins without optical energy, which is then turned on at mid-bit.

The intrinsic characteristics of optical energy and devices allow data bus architectures that are significantly different from the traditional electrical ones. Figure 2.4 shows several examples of the possibilities. Figure 2.4a is a star coupled bus in which the bus controller (BC) transmits an optical signal to the reflective coupler (basically a mirror), which reflects it to the remote terminals (RT). When an RT

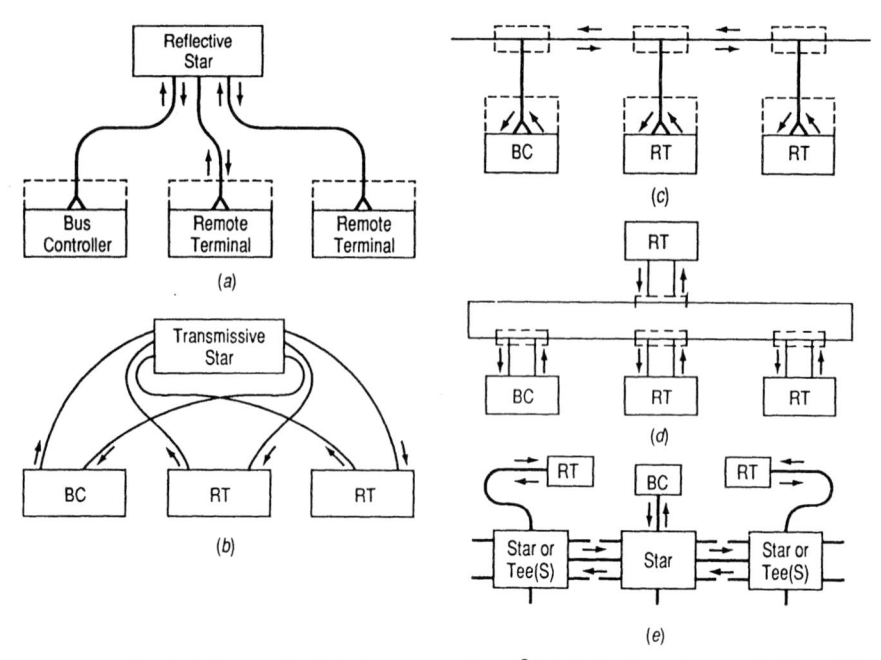

Figure 2.4 Typical fiber optic data bus connections.[2] (*a*) Reflective star-coupled bus; (*b*) bi-directional T-coupled bus; (*c*) transmissive-coupled bus; (*d*) uni-directional T-coupled bus; (*e*) bidirectional complex star or T-coupled bus.

responds, the reverse signal flow occurs. Figure 2.4*b* is a star coupler in which the transmit (output) signal lead and all other transmit signal leads are fused together and to the receive (input) signal leads from each units in the bus. Thus, when a signal is sent from any transmitter, it goes into the star coupler and then out to every receiver. Figure 2.4*c* is a simple fused coupler that allows a single LRU to tap onto the optical bus. Typically, in a fused coupler the LRU fibers are smaller than the main bus fiber, a configuration known as an asymmetric coupler.[3] Using these basic types of couplers, architectures like those shown in Figs. 2.4*d* and 2.4*e* can be constructed.

In closing the discussion on 1773, there are two caveats about optical buses, one concerning the definition of connector vis-à-vis coupler and the severe (relative to electrical) connector and coupler attenuation losses. First, an optical connector is the same as an electrical connector with the individual fibers inserted into pins just as wires are, and it is designed to be mated and demated many times; however, optical connectors are much more prone to contamination and damage by foreign material than are electrical connectors. An optical coupler is a permanent fusion of two or more fibers, even more permanent than an electrical wire solder joint, and cannot be disassembled.[3] Generally, connectors have more loss than couplers; the increased losses in connectors is the price that must be paid for a mating and demating capability. Second, the signal losses experienced in couplers and connectors are much higher than those in their electrical equivalents. These higher losses can, in some cases, reduce the bus configuration options. As optical technology improves, these losses will be reduced, allowing the optical bus designer the flexibility that is available to electrical bus designers.

High-Speed Data Bus

As data bus technology progressed and the need for exchanging data grew exponentially in modern military aircraft, the need for a much higher bandwidth data bus became apparent. In response to this need a high-speed data bus (HSDB) concept was developed and applied for the first time in the F-22 Advanced Tactical Fighter. Detailed information on the F-22 implementation of the HSDB is not available, so this discussion will focus on a generic version found in SAE Aerospace Standard (AS) 4074.1.[4] Based on available information, the F-22 version of the HSDB is apparently functionally very similar to the generic version in AS 4074.1.[5]

As described in AS4074.1, the HSDB can transmit up to 50 Mbit/s using several different bus topologies. One familiar topology is the linear bus (like MIL-STD-1553). For an optical medium implementation of the bus there are other topology options available such as the familiar star topology in which signals from each terminal are sent via a fiber to the point where the fibers from all terminals are fused together (the star) and then passively transmitted to the other terminals over a fiber from the star back to the terminals. This architecture is described in SAE Aerospace Information Report (AIR) 4288, *User's Handbook for AS4074.1 Linear Token-Passing Bus,* and is functionally identical to one presented in DOD-STD-1773.[2] Signal losses limit this topology to a maximum of 16 terminals.

Frames are the basic unit of information exchange. Figure 2.5 shows the format of the three types of frames on a HSDB. All frames begin with a Start Delimiter and end with an End Delimiter. Both delimiters are 4 bits long. The bit pattern for each delimiter depends on the specific bus protocol being implemented.

The HSDB uses distributed control in which each station* is permitted to transmit only when it has received the token frame, shown in Fig. 2.5a, that is passed to it by the previously transmitting station

---

Station is used in AS4074.1 in the same manner as *terminal* is used in other data bus standards.

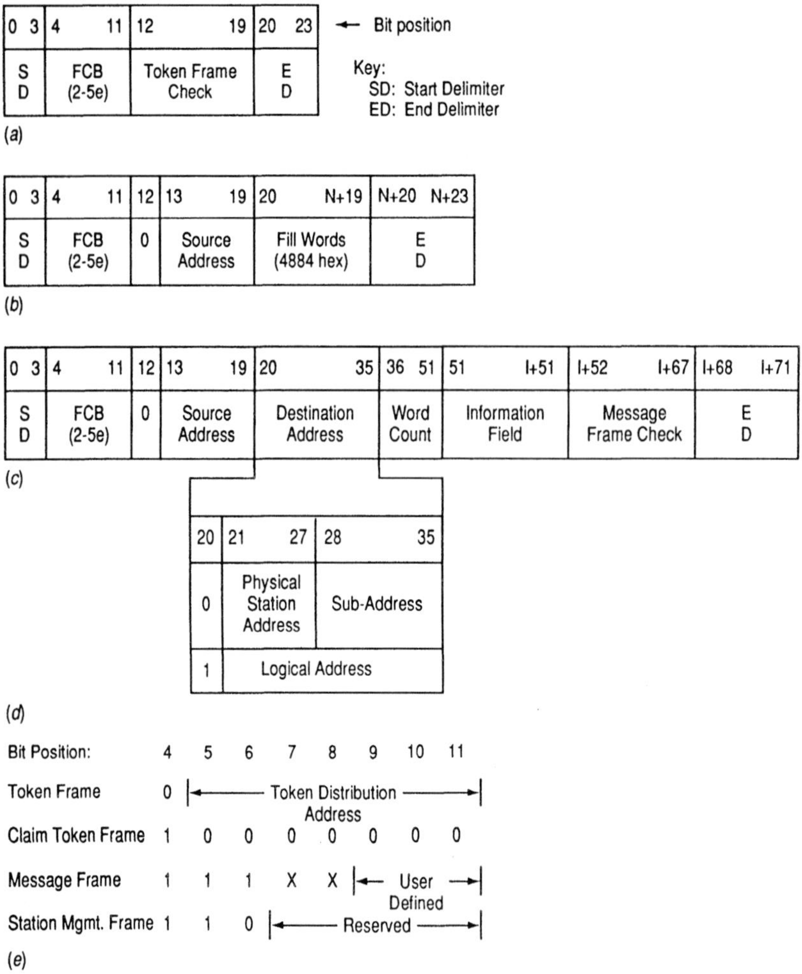

30   Chapter Two

0 3	4 11	12 19	20 23	← Bit position
SD	FCB (2-5e)	Token Frame Check	ED	Key: SD: Start Delimiter ED: End Delimiter

(a)

0 3	4 11	12	13 19	20 N+19	N+20 N+23
SD	FCB (2-5e)	0	Source Address	Fill Words (4884 hex)	ED

(b)

0 3	4 11	12	13 19	20 35	36 51	51 l+51	l+52 l+67	l+68 l+71
SD	FCB (2-5e)	0	Source Address	Destination Address	Word Count	Information Field	Message Frame Check	ED

20	21 27	28 35
0	Physical Station Address	Sub-Address
1	Logical Address	

(c)

(d)

Bit Position: 4   5   6   7   8   9   10   11
Token Frame   0 |← Token Distribution →|
                                    Address
Claim Token Frame 1   0   0   0   0   0   0
Message Frame   1   1   1   X   X |← User →|
                                              Defined
Station Mgmt. Frame 1   1   0 |← Reserved →|

(e)

Figure 2.5 HSDB frame formats.[5,6] (a) Token; (b) claim token; (c) message; (d) details on message frame format destination address (bits 20–35); (e) frame control byte (FCB) structure. Reprinted with permission. © 1991 from SAE Aerospace Information Report 4288, *Linear Token Passing Multiplex Bus User's Handbook*.

upon time out of the token-holding timer (THT). The token passes sequentially to the next higher logically numbered active station on the bus. (There may be inoperative stations or there may be gaps in the station logical number sequence to allow for expansion.) The highest logically numbered station passes the token to the lowest logically numbered station to repeat the token-passing cycle. When the bus is powered up or reset, the initial message transmission is made by the station that successfully transmits a claim token frame, Fig. 2.5*b*.

Message frames, Fig. 2.5c, are the preponderance of traffic on a HSDB. In Fig. 2.5c the destination address, bits 20 through 35, can be either a physical station address or a logical address. If it is a physical address, bit 20 is set to a logical 0, the next 7 bits contain the physical address of the station to which the message is being sent, and the last 8 bits contain the subaddress for an entity within the station. If bits 20 through 35 are a logical address, bit 20 is set to a logical 1, and the remaining bits describe an address in the message filter located in the bus interface unit (BIU). Messages can be up to 4096 words long, compared to 32 words in the case of MIL-STD-1553.

Station management frames are used to gain access to the terminal for testing and diagnosis and for updating its performance.

AS4074.1 is a generic HSDB standard. Specific implementations are described in "slash sheets" found in an appendix to the standard. The preferred implementation medium for the HSDB is optical fiber, although wire is acceptable. In a typical fiber optic medium the transmitter output power for a high signal is −2.0 ±2.0 dBm and −15 dBm for a low signal. The receiver minimum sensitivity is −32.5 dBm with an operating range of 21 dB. The optical signal wavelength should lie between 800 and 880 nm with a spectral bandwidth of 60 nm. The cable has a core diameter of 200 μm, a cladding diameter of 280 μm, and a numerical aperture of 0.2. The receiver bit error rate is 10^{-10}.[4]

In a typical wire implementation the transmitter output voltage is ±7.5 V ±20 percent, where the positive voltage is a logical 1 and the negative voltage is a logical 0. The receiver maximum input voltage is ±12 V, and the minimum input voltage is ±0.1 V. It has a dynamic range of 26 dB and a bit error rate of 10^{-12}. The characteristic impedance of the bus is 50 Ω ±20 percent.[4]

Data Buses Using ARINC 429

Just as MIL-STD-1553 is the basis for digital buses in modern military aircraft, "ARINC Specification 429 Digital Information Transfer System, Mark 33," 429 as it is commonly known, is the basis for digital buses in modern civil aircraft.[7] These two standards establish vastly different approaches to data buses that reflect the differences in design philosophy and performance priorities for military and civil aircraft. Requirements for minimum weight and maximum flexibility drove 1553 to operate at 1 Mbit on a bidirectional bus, while certification requirements drove 429 to operate at either 12 to 14.5 or 100 kbit on a simplex bus.

A simplex bus is one on which there is only one transmitter but multiple receivers (up to a maximum of 20 in the case of 429). If receipt of a message by a given sink R(n) is required by the source T, a separate bus with R(n) as the source and T as the sink is required. To those designers who focus on military systems a simplex bus may seem cumbersome, but it can be readily certified for civil aircraft. There is no BC, RT, or bus monitor as found in 1553 buses.

Communications on 429 buses use 32-bit words with odd parity. The waveform is bipolar return-to-zero with each bit lasting either 70 or 83 μs ±2.5 percent or 10 μs ±2.5 percent depending on whether the bus is low or high speed. A low-speed bus is used for general-purpose, low-criticality applications, and a high-speed bus is used for transmitting large quantities of data or flight critical information. Figure 2.6 shows the five basic word formats in 429. There are two formats for numerical data, one for discretes, and two for alphanumeric data, which are encoded using International Standards Organization (ISO) Alphabet No. 5.

Figure 2.6a shows the general form for binary coded decimal (BCD) numeric data words. All BCD and BNR (see next paragraph) data words have a five-character label assigned in 429. The first three characters are octal bits from 000 to 377 encoded binarily in bits 1 through 8. The last two characters of the label are hexadecimal and are used to identify bus sources. They are encoded hexadecimally in the two least significant digits of a BCD word with an octal 377 label. Bits 9 and 10 are a source/destination identifier (SDI). If source/destination identification is not appropriate for a particular installation, bits 9 and 10 should be set to binary zero. Data are encoded to the resolution necessary in bits 11 through 29. Bits not required to provide adequate resolution may be used for discretes or filled with binary zeros. Bit 29 is the most significant data bit. Bits 30 and 31 are the sign/status matrix, as described in Table 2.3. Bit 32 is the parity bit used to achieve odd parity.

Figure 2.6b is the format for straight binary (BNR) data or angular data which is encoded using the 2's complement fractional binary notation. For BNR coding the first 10 bits serve the same function as in BCD words. However, for BNR the data are encoded using only bits 11 through 28 where bit 28 is the most significant bit. Since this is a binary scheme, a maximum possible value for the variable being encoded must be established, and each bit of the data string is assigned an inverse power of two fractional parts of the maximum value. The receiver then totals up these fractional parts, each represented by a logical 1 in the data stream, to determine the final value up to the maximum possible. ARINC 429 contains a listing of variables that may be encoded in BNR format and their maximum permissible value. Bits 29 through 31 are the sign/status matrix for BNR words,

Avionics Systems Essentials I: Data Buses 33

32	31	30	29	28	27	26	25	24	23	22	21	20	19	18	17	16	15	14	13	12	11	10	9	8	7	6	5	4	3	2	1

| P* | SSM | MSB | Data, Pads, or Discretes | LSB | SDI | Label |

a.) Generalized BCD Word Format

| P | SSM | MSB | Data, Pads, or Discretes | LSB | SDI | Label |

b.) Generalized BNR Word Format

| P | SSM | MSB | Discretes | LSB | SDI | FltCont: 145–147, 270–276 Maint: 155–161, 350–354 |

c.) Discrete Word Format

| P | SSM | Format TBD | Word count or BNR equiv. (0's if 1 word message.) | Octal 355 |

d.) Acknowledgement: Initial Word Format

| P | 0 0 / 1 0 | Format TBD | Octal 355 |

e.) Acknowledgement: Intermediate Word Format
f.) Acknowledgement: Final Word Format

| P | 0 | 1 | ISO Alph#5 "STX" | Spares (zeros) | Word count or BNR equiv. (0's if 1 word message.) | Maint: 356; Alpha: 357 |

g.) Alphanumeric (ISO Alphabet No. 5) Data: Initial Word Format

| P | 1 | 1 | Spares (zeros) | F** | Char. size | Int. | Color | Line Count | Maint: 356; Alpha: 357 |

** = Flashing display
h.) Alphanumeric (ISO Alphabet No. 5) Data: Control Word Format

| P | 0 | 0 | Character No. 3 | Character No. 2 | Character No. 1 | Maint: 356; Alpha: 357 |

i.) Alphanumeric (ISO Alphabet No. 5) Data: Intermediate Word Format

| P | 1 | 0 | Character No. n | Character No. n−1 | Character No. n−2 | Maint: 356; Alpha: 357 |

j.) Alphanumeric (ISO Alphabet No. 5) Data: Final Word Format

* = Parity bit (odd)

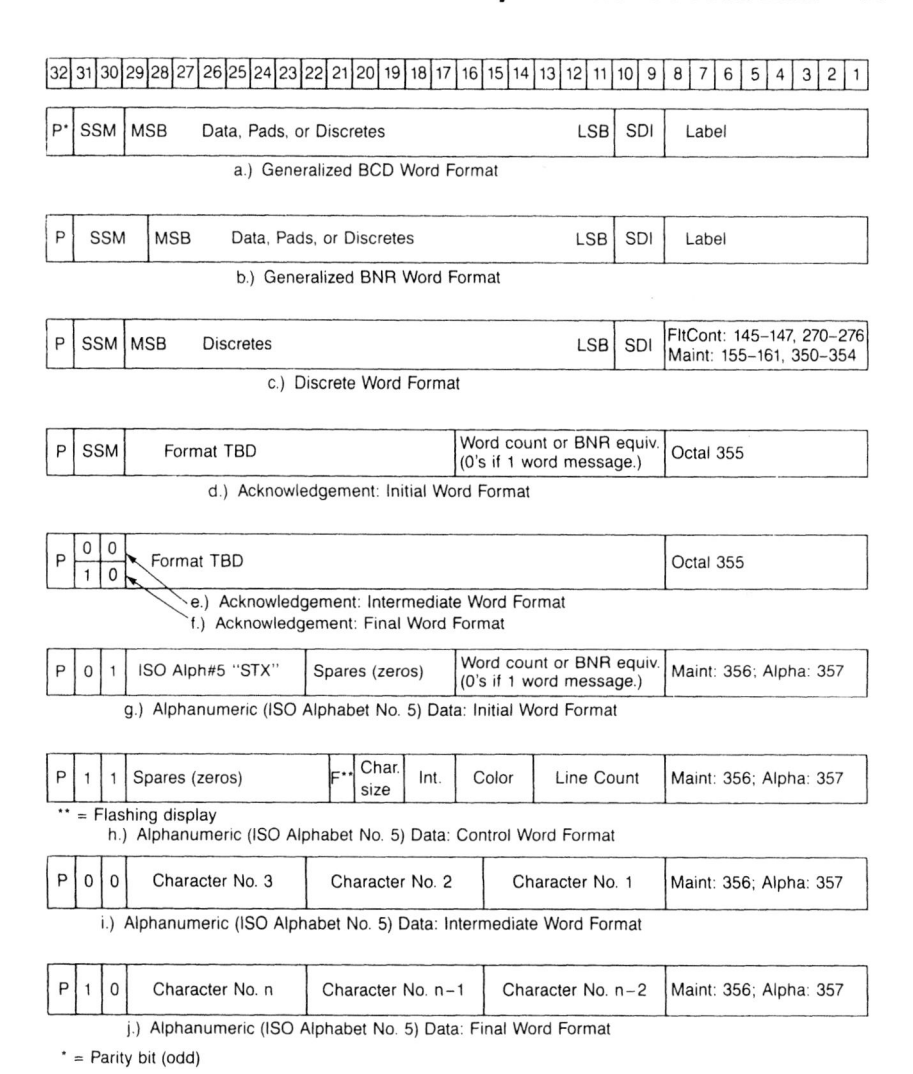

Figure 2.6 ARINC 429 BCD, BNR, discretes, acknowledgments, and alphanumeric word formats.[7]

as determined from Table 2.3. Once again, bit 32 is the parity bit for odd parity.

Figure 2.6c shows the general form of a discrete word. Note once again that bits 1 through 8 are label bits and bits 9 and 10 are for SDI. Discretes can be transmitted in bits 11 to 29. Bits 30 and 31 function the same as their counterparts in the BCD words, and bit 32 is the parity bit. There are two types of discrete words: general purpose and dedicated. General-purpose words are assigned labels octal 145

TABLE 2.3 Sign-Status Matrix.

Part A: BCD Numeric; Discrete; Acknowledgment, ISO, and Maintenance (AIM) Data; and File Transfer Words

Bit no. 31	30	BCD numeric and discrete	AIM	File transfer
0	0	Plus, north, east, right, to, above	Intermediate word	Intermediate word, plus, north, east, right, to, above
0	1	No computed data	Initial word	Initial word
1	0	Functional test	Final word	Final word
1	1	Minus, south, west, left, from, below	Control word	Intermediate word, minus, south, west, left, from, below

Part B: BNR Numeric Data Words

Bit no. 31	30	20	
0	0	0	Failure warning, plus, north, east, right, to
0	0	1	Failure warning, minus, south, west, left, from
0	1	0	No computed data, plus, north, east, right, to
0	1	1	No computed data, minus, south, west, left, from
1	0	0	Functional test, plus, north, east, right, to
1	0	1	Functional test, minus, south, west, left, from
1	1	0	Normal operation, plus, north, east, right, to
1	1	1	Normal operation, minus, south, west, left, from

through 147, 155 through 161, 270 through 276, and 350 through 354 and are used for data exchange in flight control computers and for maintenance diagnostics. Dedicated discrete words refer to those BCD and BNR words that have spare bits filled with discrete data. The discrete bits within the word will generally qualify (or describe) the functions of a specific equipment type.

General-purpose maintenance data words are formatted the same as BCD, BNR, or discrete data. Maintenance data words never contain ISO Alphabet No. 5 messages.

Acknowledgment message formats are shown in Fig. 2.6*d* through *f*. Although provision has been made for these messages, no need has yet been identified; therefore details on the structure of information in the data bits are undefined. All acknowledgment messages use octal label 355, and bits 9 through 16 of the initial word show the word count in BNR form. Bits 30 and 31 follow the sign/status matrix shown in Table 2.3 for initial, intermediate, and final words.

ISO Alphabet No. 5 messages and maintenance messages which require text follow the format shown in Fig. 2.6*g* through *j*. All messages of these types use labels octal 356 (maintenance) or 357 (ISO Alphabet No. 5) and adhere to the sign/status matrix in Table 2.3, Part A. Initial words, as shown in Fig. 2.6g, contain a word count in bits 9 through 16 in BNR code, bits 17 through 22 are spares that are set to binary zero, and bits 23 through 29 contain the ISO control character *STX*.

If the messages are being used to drive displays, the second word will be a control word with a format like the one shown in Fig. 2.6h. For control words, bits 9 through 13 contain the line count for the display, bits 14 through 16 set the color, bits 17 and 18 set the intensity, bits 19 and 20 specify the character size, and bit 21 is set to logical 1 if the display is to flash. Bits 22 through 29 are spares that are set to binary 0.

Intermediate and final words conform to the formats shown in Fig. 2.6i and j. For both types of words three ISO characters are encoded in bits 9 through 29. If there are less than three characters in the final word, the bits corresponding to the unused character(s) are set to binary zero.

File data transfer is a command/response protocol for transferring large amounts of data in cases such as flight management computer programs and flight plan loading and update. Communication is usually at the 100-kbit rate. A file may contain up to 127 records, which, in turn, may each contain up to 125 data words. These words may be either BNR numeric or ISO Alphabet No. 5. There are three word types: initial, intermediate, and final. Bits 30 and 31 are used in accordance with Table 2.3, Part A, to identify the word type. Initial words may be further characterized as one of eight types: Request to Send, Clear to Send, Data Follows, Data Received OK, Data Received Not OK, Synchronization Lost, Header Information, and Poll. Figure 2.7 shows the detailed bit pattern of each initial word type. A symbolic typical successful transfer of data is shown below:

	Source to sink (bus 1)	Sink to source (bus 2)
t1	Initial word: Request to Send	
t2		Initial word: Clear to Send
t3	Initial word: Data Follows	
tn − 1	Intermediate word	
tn	Final word	
tn + 1		Initial word: Data Received OK

The Header Information and the Poll initial words require additional comment. The header information word allows a transmitter to send file size information without any commitment to actually transfer the file itself. The poll word is used for handshaking between two terminals.

36 Chapter Two

32	31	30	29 28 27 26 25 24 23	22 21 20 19 18 17 16	15 14 13 12 11 10 9	8 7 6 5 4 3 2 1
P*	0	1	ISO Alph#5 "DC2"	Blank (zeros)	No. of records to be sent (BNR) (<=127)	File Label

a.) Request to Send (Transmitter to Receiver)

P	0	1	ISO Alph#5 "DC3"	Blank (zeros)	(See Note 1)	File Label

(See Note 2)

Note 1: Bits 9–15: "0000000" if receiver is not ready to accept data
If receiver is ready then:
BNR count of number of maximum length records or
Number of 32-bit words receiver can accept

Note 2: Bit 22: "0" when receiver is not ready to accept data and
when bits 9–15 are maximum length record count
"1" when bits 9–15 are 32-bit word count

b.) Clear to Send (Receiver to Transmitter)

P	0	1	ISO Alph#5 "STX"	Rcd. Seq. No. (BNR)	No. of words in record (BNR)	File Label

c.) Data Follows (Transmitter to Receiver)

P	0	1	ISO Alph#5 "ACK"	Rcd. Seq. No. (BNR)	No. of words in record (BNR)	File Label

d.) Data Received OK (Receiver to Transmitter)

P	0	1	ISO Alph#5 "NAK"	Rcd. Seq. No. (BNR) which has error	No. of words in record (BNR)	File Label

e.) Data Received Not OK (Receiver to Transmitter)

P	0	1	ISO Alph#5 "SYN"	Blank (zeros)	Blank (zeros)	File Label

f.) Synchronization Lost (Receiver to Transmitter)

P	0	1	ISO Alph#5 "SOH"	Binary zeros	No. of records to be Xfrd. (BNR) (<=127)	File Label

g.) Header Information (Transmitter to Receiver)

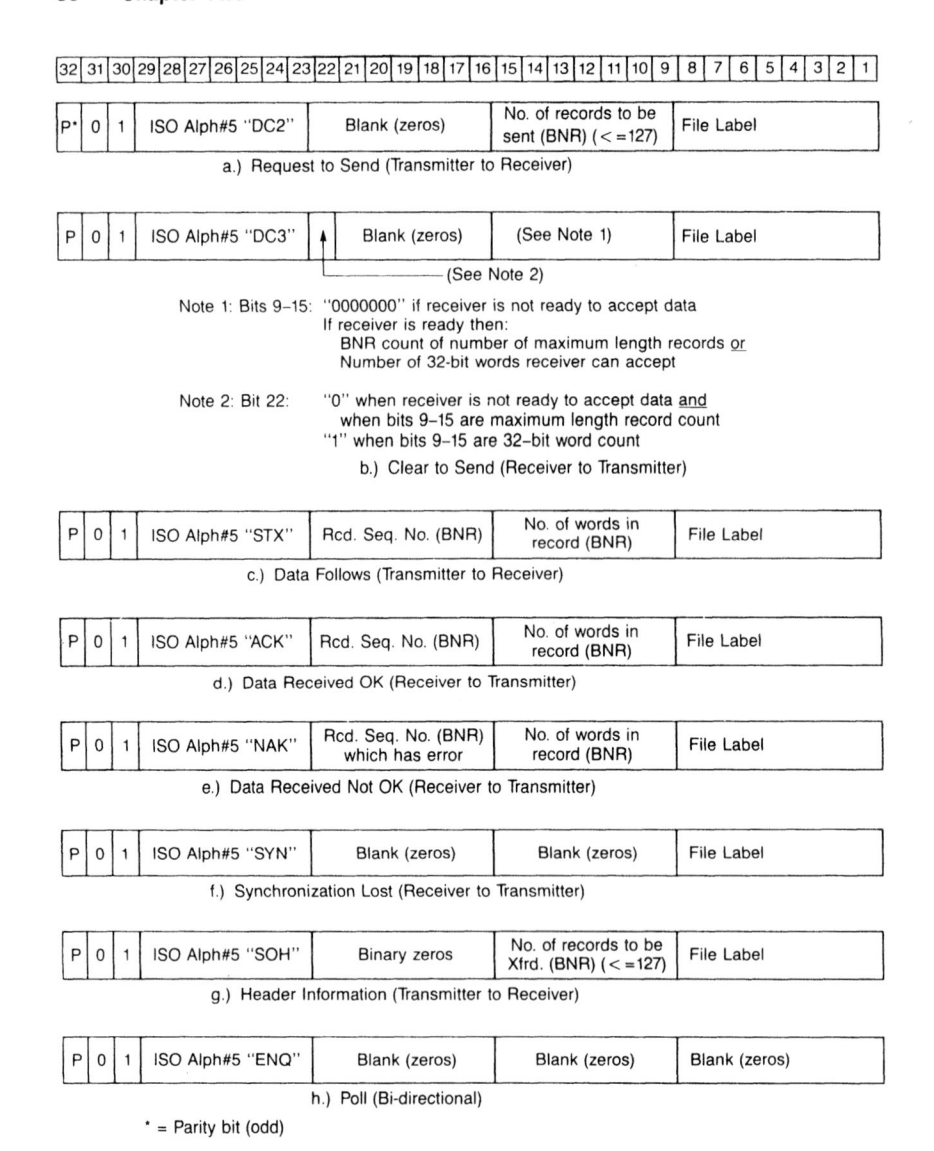

P	0	1	ISO Alph#5 "ENQ"	Blank (zeros)	Blank (zeros)	Blank (zeros)

h.) Poll (Bi-directional)

* = Parity bit (odd)

Figure 2.7 ARINC 429 file data transfer initial word types.[7]

Synchronization is achieved on each word by sensing the transition on the first bit. A minimum of 4 bit times is inserted between successive word transmissions.

ARINC 429 imposes relatively modest and readily achievable performance demands on the hardware. Figure 2.8 is a general schematic of a 429 bus. The transmitter output impedance should be in the range of 75 to 85 Ω, equally divided between the two leads. The output volt-

Avionics Systems Essentials I: Data Buses

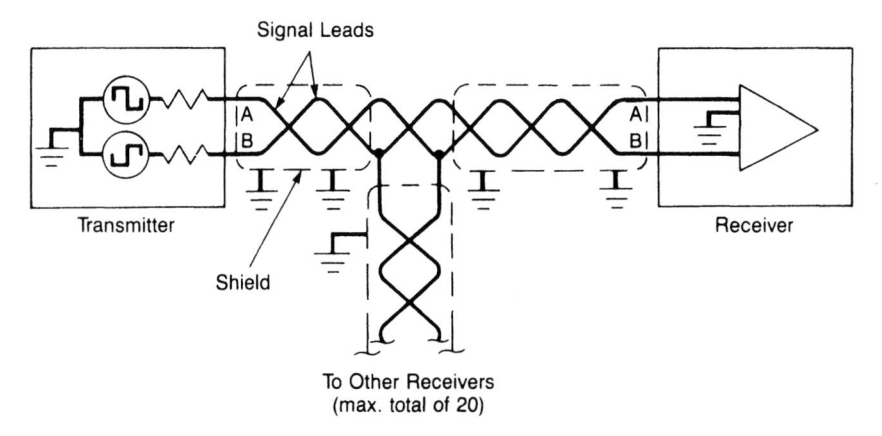

Figure 2.8 Generalized ARINC 429 bus.[8]

age V_o is ±10 ±1.0 V generated by imposing equal but opposite polarity voltages on the two leads. The null voltage is 0 ±0.5 V. For the receiver, the input resistance shall be greater than 12,000 Ω, and the input differential capacitance and the capacitance to ground shall, in both cases, be less than 50 pF. The 12,000-Ω minimum input resistance ensures that up to 20 receivers can be on the bus without overloading it and minimizes receiver interaction under fault conditions. In order to preclude continued receiver operation in a lead-to-ground fault condition, 429 has established the range of acceptable receiver voltage levels to be +6.5 to 13.0 V and −6.5 to −13.0 V and null levels from +2.5 to −2.5 V. Any signals falling outside of these levels will be ignored. Also note that a lead-to-ground fault will produce a differential voltage swing up ±5.5 V. Figure 2.9 shows the waveforms required by 429 and the permissible levels for transmitter and receiver voltages.

The cable used in 429 buses is a twisted, shielded pair of 20- to 26-gauge conductors. The shield is grounded at both ends of the cable run and at all production breaks. Although there is no specification placed on the cable impedance, it generally falls in the range of 60 to 80 Ω.

Data Buses Using ARINC 629

Just as MIL-STD-1553 has been supplemented by DOD-STD-1773 and the HSDB to bring newer technology to military aircraft data bus applications, so too has ARINC 429 been supplemented by ARINC Characteristic 629 Multi-Transmitter Data Bus to increase the capability of civil aircraft data buses.[8] ARINC 629 uses word formats that are very similar to those in MIL-STD-1553, but it does not have a bus controller, long felt by civil avionics manufacturers and regulatory agen-

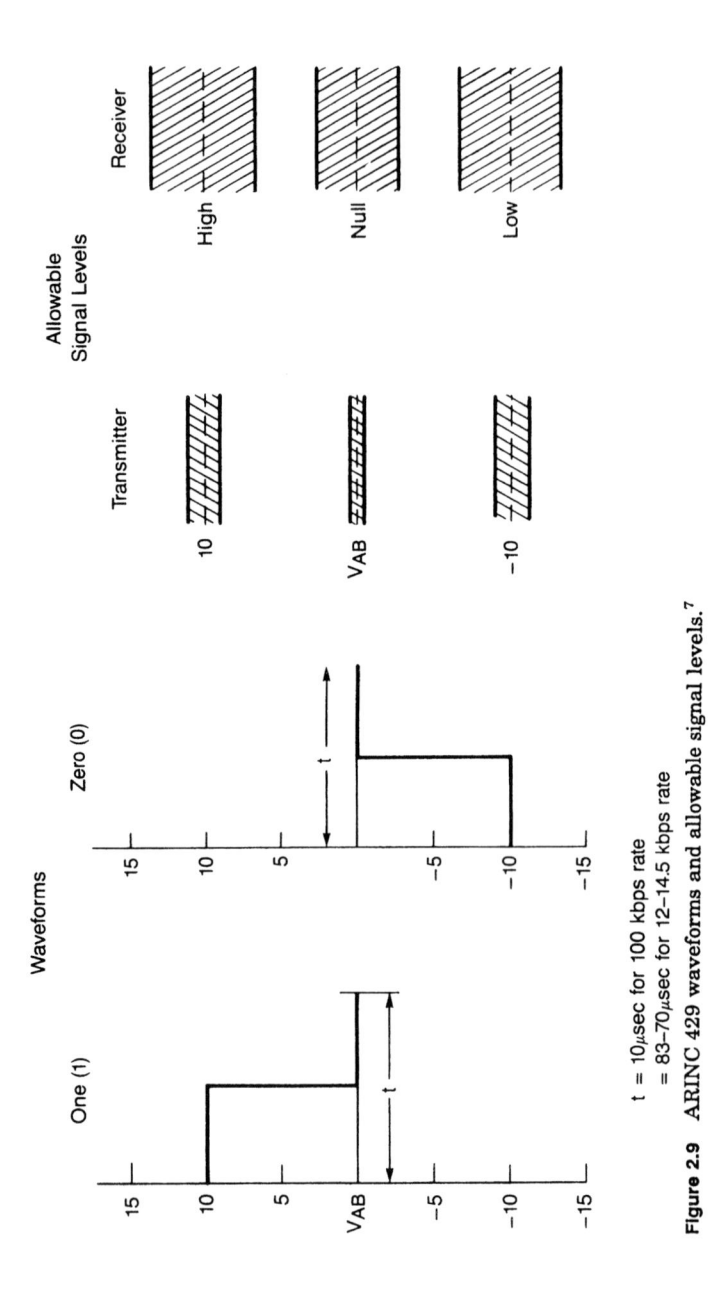

Figure 2.9 ARINC 429 waveforms and allowable signal levels.[7]

cies to be the weakest point in a bus and a potential major source of catastrophic failures. Instead, ARINC 629 gives each terminal autonomous access to the bus based upon meeting three timing conditions stored in a Transmit Personality programmable read-only memory, or PROM (XPP) in the host terminal. One of these timing conditions is unique to the terminal. Another difference from MIL-STD-1553 is the use of a Receive Personality PROM (RPP) to identify the labels of messages to be recorded from the bus.

Figure 2.10 shows the ARINC 629 word formats. The words are 20 bit times long with a 3-bit synchronization interval at the beginning of each word and a parity bit (odd) at the end. Notice the label word, Fig. 2.10a, is characterized by a synchronization pattern that starts in a high state and transitions to a low state in the middle of the second bit time. All other words follow the reverse pattern; thus, the synchronization pattern alone is sufficient to identify a label word. For label words the 16 bits between the synchronization pattern and the parity bit are designated as label bits (LBs). Notice the label bits are numbered in reverse order from the basic word bit positions (i.e., LB15 is in bit time 4 and LB0 is in bit time 19). For data words the 16 bits between the synchronization pattern and the parity bit are designated as data bits (DB), where DB15 is in bit time 4 and DB0 is in bit time 19.

Messages are composed of a series of wordstrings. The first word of any wordstring is the label word with its unique synchronization pattern followed by the source channel identification (CID), the wordstring label, and the parity bit. CID is used to identify a unique unit when there are two or more identical, redundant units on the same bus such as triple redundant inertial reference units.

The first data word in any message should be a system status word, Fig. 2.10b. The system status word shows the health and operating environment of the originating system. Bit 15 is the master failure warning flag which, if set to a logic 1, signifies that the system has had a complete failure and all data from that system are bad. Bit 14 is a partial failure flag and some data may be bad. Bit 13 is a special case of bit 14 in which the system cannot compute reliable data for reasons other than system failure. If either bit 14 or 13 is set to a logical 1, the receiving subsystem should watch for an invalid data code in the parameter validity word later in the message.

For complex systems a function status word, Fig. 2.10c, is used to describe the condition of specific functions in the originating system. The function status word has a tree structure that facilitates retrieving detailed information on the function(s). If any function reported in the function status word is defective, DB15 is set to a logical 1. DB14 through DB12 are assigned to specific functions and if any of those bits are set to a logical 1, the function represented by that bit is de-

40 Chapter Two

| 1 | 2 | 3 | 4 | 5 | 6 | 7 | 8 | 9 | 10 | 11 | 12 | 13 | 14 | 15 | 16 | 17 | 18 | 19 | 20 |

| | LB | 15 | 14 | 13 | 12 | 11 | 10 | 09 | 08 | 07 | 06 | 05 | 04 | 03 | 02 | 01 | 00 | |
| H-L SYNC | CID | Label | P |

LB: Label Bit   H-L SYNC: High-Low Synchronization Pattern
CID: Channel Identification   P: Parity

(a)

| | DB | 15 | 14 | 13 | 12 | 11 | 10 | 09 | 08 | 07 | 06 | 05 | 04 | 03 | 02 | 01 | 00 | |
| L-H SYNC | See text | Spare | Word count or pad | P |

DB: Data Bit   L-H SYNC: Low-High Synchronization Pattern

(b)

| | DB | 15 | 14 | 13 | 12 | 11 | 10 | 09 | 08 | 07 | 06 | 05 | 04 | 03 | 02 | 01 | 00 |
| L-H SYNC | H | Function | Function Subgroups | P |

H: Health Bit (See text)

(c)

| | DB | 15 | 14 | 13 | 12 | 11 | 10 | 09 | 08 | 07 | 06 | 05 | 04 | 03 | 02 | 01 | 00 |
| L-H SYNC | Refresh Ctr | Parameter Validity Bits | P |

(d)

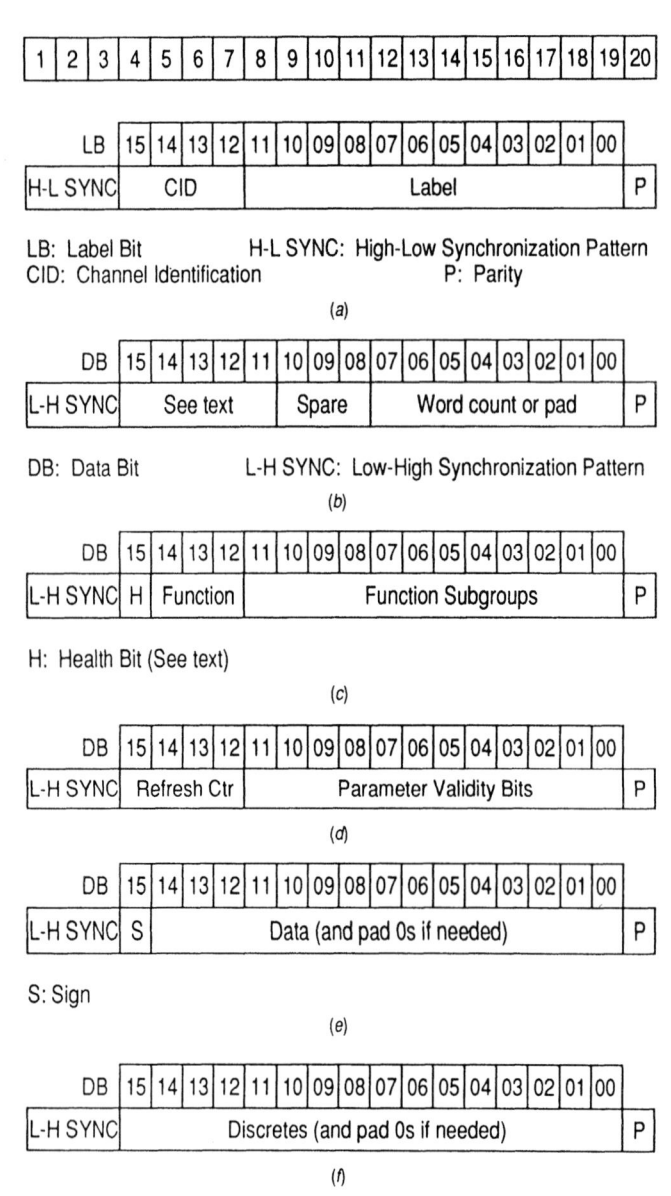

| | DB | 15 | 14 | 13 | 12 | 11 | 10 | 09 | 08 | 07 | 06 | 05 | 04 | 03 | 02 | 01 | 00 |
| L-H SYNC | S | Data (and pad 0s if needed) | P |

S: Sign

(e)

| | DB | 15 | 14 | 13 | 12 | 11 | 10 | 09 | 08 | 07 | 06 | 05 | 04 | 03 | 02 | 01 | 00 |
| L-H SYNC | Discretes (and pad 0s if needed) | P |

(f)

Figure 2.10 ARINC 629 word formats.[8] (a) Generalized label word format; (b) system status word; (c) function status word; (d) parameter validity word; (e) generalized binary (BNR) data word; (f) discrete word.

fective. If the function assigned to DB14 is defective, details about the defect are provided in DB11 through DB08. Details on the function assigned to DB13 are given in DB07 through DB04 and DB03 through DB00 describe the function assigned to DB12. Thus the tree structure allows the data recipient to explore as far as necessary to understand the details of the function defect. For functions that only need the data as an input, details on the defect are unimportant; it is only necessary to know that the data may be unusable. A maintenance and diagnostic system would examine all levels of the tree for complete details about the defect.

There are two protocol schemes for ARINC 629: basic or combined. In the basic protocol the bus operates in either the periodic or aperiodic mode but not simultaneously. The basic protocol is recommended for high-integrity applications, such as flight controls, where there will be only occasional long messages that will drive the bus from the preferred periodic mode into the aperiodic mode. For those applications where there will be more frequent long messages, such as a flight management system, the combined protocol with both periodic and aperiodic traffic is preferred. In the combined protocol the periodic messages have priority and are guaranteed to be transmitted on a set schedule (see the later discussion on bus access protocol), while the aperiodic data are transmitted only when the bus is not transmitting periodic data. The aperiodic data are assigned to one of two priority levels: Level 2 or Level 3. (Periodic data are Level 1.) All Level 2 data are transmitted before Level 3.

The unique feature of ARINC 629 is that access to the bus to transmit by a given terminal is based on meeting three timing conditions, one of which in unique to the terminal. Thus, there is no need for a bus controller, and bus access is autonomously determined by each terminal.* Figure 2.11 shows the time-based operation of ARINC 629 in the periodic mode. The three timing conditions are designated by transmit interval (TI), synchronization gap (SG), and terminal gap (TG). The TI is selected to be compatible with the maximum update rate for any parameter on the bus. TI can range from 0.5 to 64 ms and applies to all terminals on the bus. The SG is the second longest time on the bus and must be one of four values, 17.6875, 33.6875, 65.6875, or 128.6875 μs, for all terminals. The SG should be as short as possible to minimize dead time on the bus. The TG is unique to a given terminal and ranges from 1 to 125 μs. All three time intervals are assigned by the airframe manufacturer.

The timing conditions determine the bus operation as follows (refer

---

*In recognition of this autonomous concept, when ARINC 629 was originally developed by Boeing Commercial Airplanes, it was known as DATAC, Digital Autonomous Terminal Access Communication.

Figure 2.11 ARINC 629 timing diagram.[8]

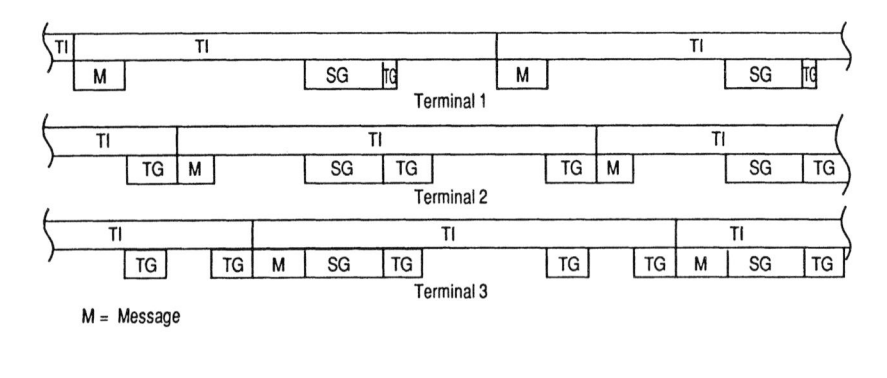

M = Message

TI = Transmit Interval
SG = Synchronization Gap
TG = Terminal Gap

Notes:
Not To Scale.
Only Elapsed Timers Are Shown.

to Fig. 2.11, which shows an example for three terminals). The TI for Terminal 1 has clocked out and the bus is quiet. Therefore, Terminal 1 sends its message and restarts its TI clock. Since Terminals 2 and 3 have not clocked out their TIs, they restart their TG timers upon completion of the Terminal 1 message. When Terminal 2 TI and TG have clocked out and the bus is quiet, it sends its message and restarts its TI timer. Terminal 3 restarts its TG timer upon completion of the Terminal 2 message. When the Terminal 3 TG and TI have clocked out, it transmits its message and restarts its TI timer. When there is no traffic on the bus, all terminals clock out the SG followed by their respective TGs. Since Terminal 1 has clocked out its SG and TG, when its TI is complete, it sends another message and the entire cycle repeats.

ARINC 629 operates at 2 Mbit/s and can be implemented in three media: wire, either inductive or voltage coupling, and optical fiber. The most developed media is an unshielded twisted wire pair with inductive coupling, which is used on the Boeing B-777. Like MIL-STD-1553, voltage coupling is not preferred because a short in any terminal could cause the entire bus to fail. Optical media are under development but are not yet in production. A current mode coupled bus can be up to 100 meters (m) long with stubs up to 40 m long. The bus has a characteristic impedance Z_o of 130 Ω ±5 percent terminated in a resistor of Z_o ±5 percent. The stubs have a characteristic impedance of 100 Ω ±5 percent. The bus cable is #20 AWG. The stub cable is a six-conductor #22 AWG configured in two individually twisted shielded pairs with separate power return conductors inside but insulated from the shield. Figure 2.12 shows a typical inductive bus coupler.

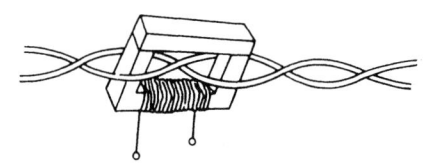

- U-I Core Implementation of Remotely Located Coupler

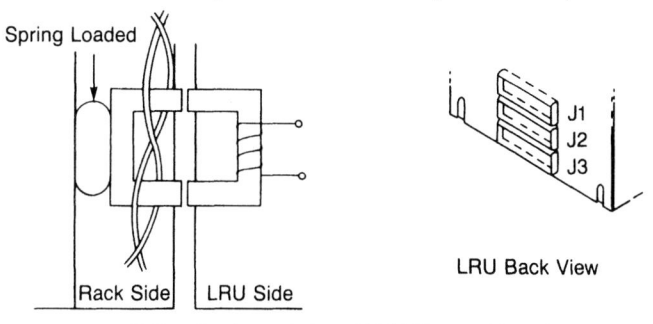

- C Core Implementation of LRU Mounted Coupler

Figure 2.12 ARINC 629 inductive mode coupler.[8]

The optical implementation of ARINC 629 offers the same alternative bus configurations as those of DOD-STD-1773, discussed earlier. The designer is encouraged to examine these alternatives to determine if they offer advantages over the conventional bus configurations for the particular application at hand.

ARINC 659 Is the Civil Avionics Backplane Data Bus

The Boeing B-777 Airplane Information Management System (AIMS) use of the ARINC 659 backplane data bus for intracabinet data transfer among the AIMS line replaceable modules (LRMs) is the first application of 659.*[9] (See Chap. 7 for a discussion of AIMS.) ARINC 659, originally developed by Honeywell, Inc., and called SAFEbus,† is a table-driven, proportional-access bus in which all of the bus activity is controlled by a table memory contained in each LRM. Like ARINC 629, there is no central bus controller.

ARINC 659, Fig. 2.13, is a linear, half-duplex, serial data bus in a dual-dual configuration. Each LRM contains identical dual redundant BIUs which are connected to dual redundant buses. Each of the dual bus pairs, A and B, has an x and a y bus. Each bus, Ax, Ay, Bx, By,

---

*Intercabinet data exchange and cabinet input-output communications are generally over ARINC 629 data buses. In cases of older design equipment, ARINC 429 may be used.

†SAFEbus is a registered trademark of Honeywell, Inc.

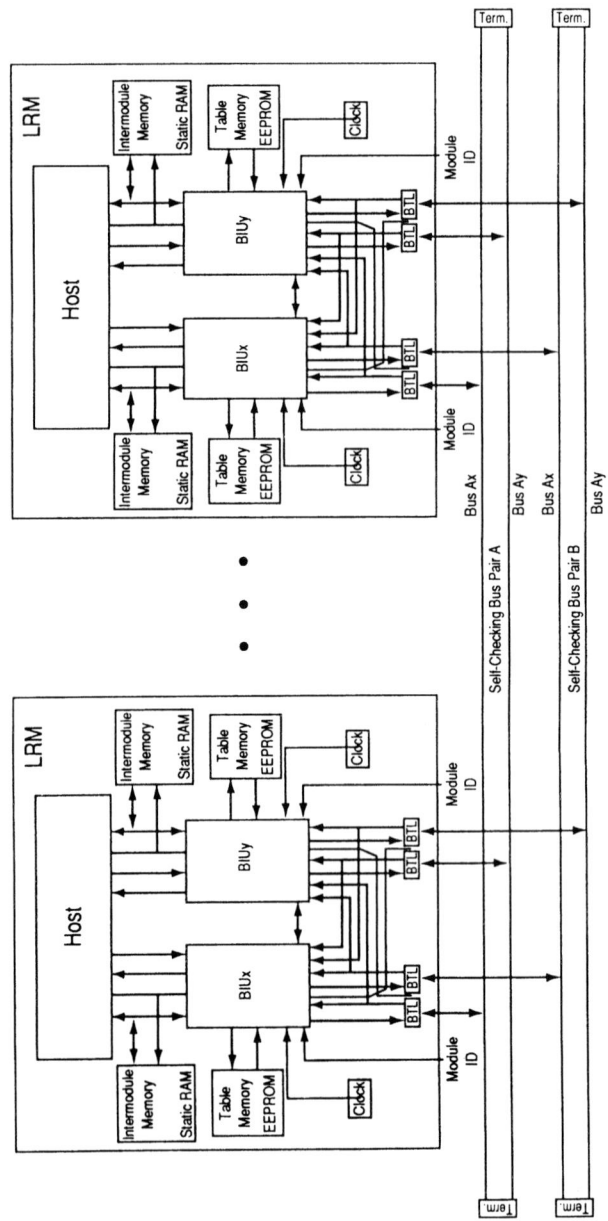

Figure 2.13 Example ARINC 659 data bus. (BTL = backplane transceiver logic)[9] (*Courtesy of Honeywell, Inc.*)

has separate data and clock bus lines for a total of eight bus lines (16 wires) in a complete installation. Bus traffic received by the LRM is subject to four comparisons: Ax:Ay, Bx:By, Ax:By, and Bx:Ay. Ax:Bx and Ay:By are not valid comparisons since they come from the same source and may have correlated errors. Thus, the bus has effectively quadruple redundancy.

The structure of 659 bus traffic is based hierarchically on frames, windows, messages, and words. A frame is a sequence of bus activity comprising windows and idle time that is cyclically repeated. A window is a time interval equal to the message transmission time (a linear function of the number of words) plus a 2-bit-time synchronization gap. A message contains from 1 to 256 32-bit words, and there are no time gaps between words, only the 2 bit times at the end of the message.

In real time avionics applications the frame rate, the inverse of the frame time, is set to provide the maximum update rate required by computations that use parameters or information received from the bus. Thus, the maximum allowable frame time is automatically established by the frame rate.

Both BIUs in an LRM have an identical PROM table memory containing the sequence and duration of every window in each frame and specific commands to transmit, receive, or be idle in a window. Figure 2.14 is a typical table memory in which the pointer shows that during the current window BIUs 3 and 3' are transmitting parameter δ_F.[10] In the next window BIUs 1 and 1' will transmit parameter N_Z. If the LRM connected to the BIUs needs the information or parameter being

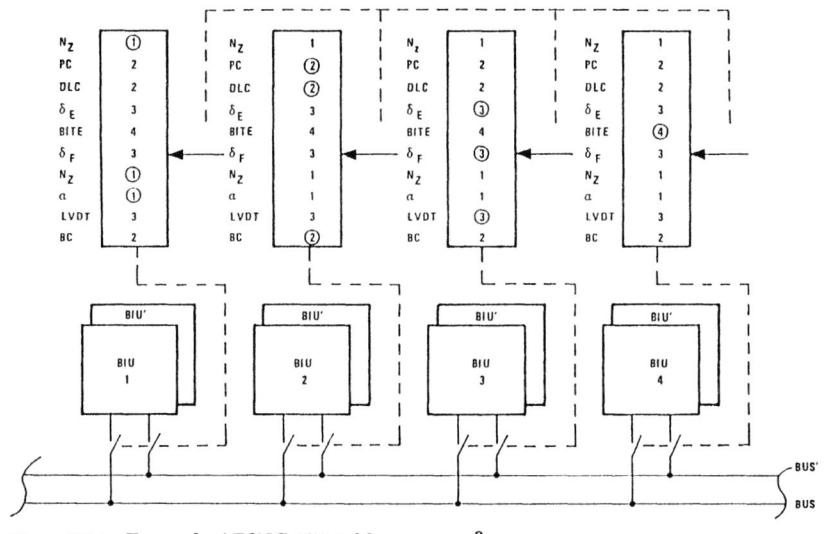

Figure 2.14 Example ARINC 659 table memory.[9]

transmitted in a window, the table memory will instruct its host BIU to be in the receive mode. If the LRM is neither transmitting nor receiving information in a window, the table memory will instruct the BIUs to be idle.

The fixed window times contribute to the overall fault tolerance of the system. If a BIU does not transmit within its designated window, all BIUs still advance the pointer to the next location at the end of the window. In order to ensure the maximum probability that a message will be sent, a master/shadow/shadow concept is used (assuming three LRUs can transmit the data). If the first LRU (master) cannot transmit because it has no fresh data at the beginning of the window, the second LRU (first slave) will wait a predetermined interval after the window opens and then will begin transmitting if it has fresh data available. If the first slave cannot transmit, the third LRU (second slave) will wait an additional interval equal to that of the second LRU and then will begin transmitting if it has fresh data available.

The table memory is the key to determinism in bus operations and, thus, ultimately the key to certifying the bus. Since all bus traffic is controlled by the easily analyzable table memory, the behavior of the bus in all possible failures modes can be determined.

ARINC 659 operates at 30 Mbit/s over either twisted wire pairs or fiber optic media. Each word is 32 bits long and is transmitted using little endian organization, that is DB0 is the least significant bit and is transmitted first. DB31, the last DB transmitted, is the sign bit. The bus can transmit all types of data including binary, alphanumeric, discrete, extended binary, real numbers, byte, and graphic data. Alphanumeric data is encoded using ISO Alphabet No. 5, as in ARINC 429.

For the wire version of 659 a logic 1 is a signal line between 1.6275 and 2.1 V ±2 percent, and a logic 0 is a signal line between 0.75 and 1.4725 V ±2 percent. Logic 1 is the released state; logic 0 is the asserted state. Each end of each bus line is terminated in a resistance of 39 Ω ±1 percent.

No signal levels have been established for the optical version of 659.

References

1. MIL-STD-1553B Digital Time Division Command/Response Multiplex Data Bus, Notice 2, 8 September 1986.
2. DOD-STD-1773 Fiber Optics Mechanization of an Aircraft Internal Time Division Command/Response Multiplex Data Bus, Notice 1, 2 October 1989.
3. Miller, Michael B., "Linear Fiber Optic Data Bus for Aircraft Applications," *SPIE Proceedings*, vol. 840, 1987, pp. 128–135.
4. Aerospace Standard AS4074.1 Linear Token Passing Multiplex Data Bus, SAE, 1988.

5. Ulhorn, Roger W., "The Fiber-Optic High-Speed Data Bus for a New Generation of Military Aircraft," *IEEE LCS,* vol. 2, no. 1, 1991, pp. 36–45.
6. Aerospace Information Report 4288, *Linear Token Passing Multiplex Data Bus User's Handbook,* SAE, 1991.
7. ARINC Specification 429: Mark 33 Digital Information Transfer System, Aeronautical Radio, Inc., 1977.
8. ARINC Characteristic 629: Multi-Transmitter Data Bus, Aeronautical Radio, Inc., November 1989.
9. ARINC Characteristic 659: Backplane Data Bus for Integrated Modular Avionics (Draft), Aeronautical Radio, Inc., May 1991.
10. Driscoll, Kevin: "Multi-Microprocessor Flight Control System, 1982," Proceedings of the IEEE/AIAA Fifth Digital Avionics Systems Conference, pp. 11.3.1–7, CH1839-0, 1983.

Bibliography

DOD-STD-1773 Fiber Optics Mechanization of an Aircraft Internal Time Division Command/Response Multiplex Data Bus, Notice 1, 2 October 1989.

Glass, M., "The 1553 Databus: Still Emerging," *Avionics,* vol. 16, no. 2, February 1992, pp. 22–27.

MIL-STD-1553B Digital Time Division Command/Response Multiplex Data Bus, Notice 2, 8 September 1986.

"Multiplex Applications Handbook," U.S. Air Force Systems Command, 1982. Available from SCI Systems, Inc., Huntsville, Alabama.

Chapter

3

Avionics Systems Essentials II: Crew Interfaces

The crew generally has only three questions about an aircraft: How well does it handle, how friendly is the cockpit, and can you get home safely in the event of trouble? This chapter examines the second of these questions—how friendly is the cockpit?

How much a crew likes an aircraft is determined, to a great degree, by how friendly the cockpit is. Faced with a confusing array of controls and displays that are difficult to see and to operate, the crew will not like the aircraft no matter how smoothly it flies or how many failures it can survive.

The microelectronics revolution brought with it not only 1s and 0s and wondrous things to do with them but also step function advances in displays and controls. Electromechanical indicators and single-purpose toggle switches were, for the most part, pushed aside by cathode-ray tubes (CRTs), flat panel displays, and multifunction control/display units.

This chapter examines the state of the art and presents performance data for the most commonly used electronic displays including CRTs, liquid crystal and luminous flat panels, and helmet-mounted, head-up, head-level, and head-down displays. Input-output devices, which also play an important role in the crew interface with the aircraft, are also examined.

Included in this chapter is guidance on laying out the cockpit using design methods such as timelines and workload analysis. Real aircraft cockpits are cited as examples of good designs.

Voice is emerging as an extremely promising tool for use in cockpits. The fundamentals and caveats for voice interactive systems are presented to enable the reader to begin using this rapidly emerging technology.

The final section in this chapter reviews some of the commonly used terms in display technology.

CRTs Are Versatile Display Devices

Cathode-ray tubes are the cornerstones of displays in all types of modern aircraft. The versatility and other performance advantages of CRTs when compared to the traditional electromechanical indicators have resulted in flexible, CRT-based displays becoming the centerpiece of every contemporary cockpit and a strong selling point for the aircraft.

The versatility of CRTs completely overshadows the sharply limited functions of traditional electromechanical indicators. In fact, the potential for CRT-based displays is limited only by the imagination of the display designer and the capabilities of the display generator. (This virtually unconstrained flexibility can sometimes open the door to needlessly complex displays.)

The reliability of CRTs in aircraft applications has been excellent. There have been no reports of catastrophic implosions that were so feared by skeptical pilots in their initial appraisal of CRTs. A CRT generally requires minimal maintenance skills: Either it has failed or it has not. There is no need for watchmaker skills like those required to repair electromechanical indicators. Maintenance personnel who service the rest of the digital avionics can readily service CRTs.

Since CRTs are active displays, that is, emitting light rather than simply reflecting ambient light, provisions must be made to dim the CRT during night flying. Because of the dark adaptability of the eye, displays that are comfortably bright in full sunlight appear as giant, blinding searchlights in a cockpit at night. Great care must be taken to ensure that the CRTs can be adequately dimmed so that the crew does not have an image "seared" onto their retinas following an extended night flight.

Monochrome displays find widespread application in military aircraft cockpits. The F/A-18 uses four monochrome CRTs including one for the head-up display, while the F-16 uses five, one of which is dedicated to the head-up display. Monochrome CRTs are well suited to fighter aircraft because of their ruggedness and high brightness, ranging from 500 to 3000 footcandles (fc). This type of CRT typically operates at about 23-kV accelerating voltage and uses a laminar flow or dispenser cathode to achieve a resolution of 200 lines per inch. The relatively high brightness is achieved through an optimum match of the CRT phosphor and an optical bandpass filter covering the CRT face. Although the filter attenuates the light emitted by the phosphor, this attenuation is more than compensated for by the filter absorbing all other wavelengths of ambient light, thereby providing a high-contrast display.

Color CRTs are becoming ubiquitous in the cockpits of modern transports, such as the MD-80, MD-11, B-757, B-767, B-777, A-310, A-320, A-330, and A-340 and in business aircraft such as the Cessna Starship. To a limited extent they are beginning to be used in military high-performance aircraft such as the F-15E, Eurofighter Aircraft, and the Dassault Rafale. These military applications of color CRTs are made possible by the development of a more rugged design and the use of shock mounts for vibration isolation. Some current-generation CRTs use a taut mask design to minimize vibration-related problems and color changes at high operating temperatures, especially at the edge of the display, caused by shadow mask thermally induced distortion. (There are other color CRT designs, which are discussed later, but the performance of these CRTs to date does not match that of the shadow masks.)

Shadow mask CRTs can operate in ambient light levels up to 8000 fc, which corresponds to the brightness of sunlight at typical aircraft-operating altitudes. Shadow mask color CRTs are not as bright as monochrome CRTs because of lower phosphor efficiency and screen blocking by the mask. However, brightness levels up to 6000 fc can be achieved. This brightness, coupled with filters similar to those used in monochrome CRTs (except designed to pass different wavelengths), produce a display that is readable even when full sunlight is striking the CRT face. Typical resolution is 84 red-green-blue (RGB) triads per inch.

All shadow mask CRTs, as the name suggests, use a thin metallic mask to ensure the proper alignment of each of three electron beams with its matching phosphor. Figure 3.1 shows the delta gun arrangement of source, shadow mask, and phosphor dots as used in modern transport aircraft display CRTs.

Although there is almost an infinite number of colors possible by mixing the three primary colors, early research on CRTs for cockpits has narrowed the practical number of colors to seven: green, red, amber, cyan, magenta, purple, and white. Table 3.1 shows the percentage of luminance of each primary color that is required to achieve the desired color.[1]

Beam penetration CRTs offer an attractive alternative to shadow mask CRTs in that they do not require a shadow mask to provide colors. Instead, an electron beam from a single gun excites the phosphor coating, which contains two layers. The outermost layer (toward the impinging beam) is a red phosphor that is excited by low-energy (<10

52   Chapter Three

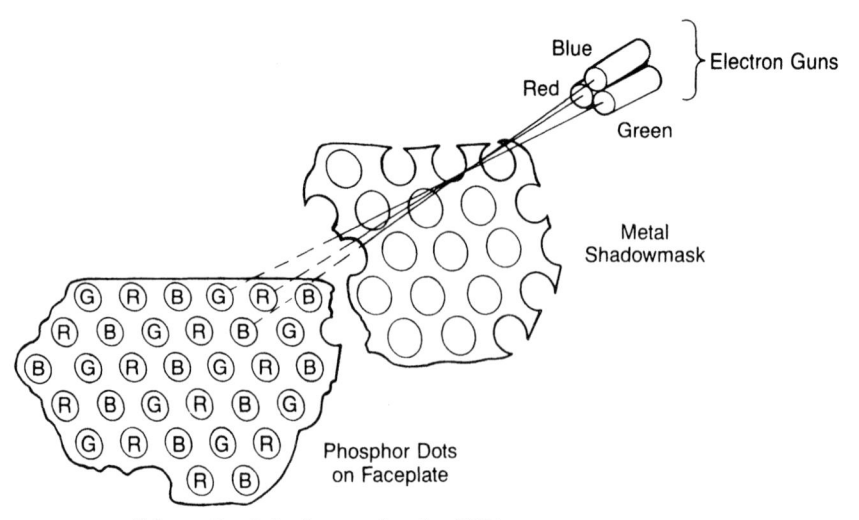

Figure 3.1 Schematic of shadow mask color CRT.

TABLE 3.1 Typical Percentage of Luminance of Primary Color Required to Produce Desired Color[1]

Color	Primary	Primary luminance (%)
Green	Green (G)	100
	Red (R)	0
	Blue (B)	0
Red	G	0
	R	100
	B	0
Amber	G	83.3
	R	88.8
	B	0
Cyan	G	64.0
	R	0
	B	100
Magenta	G	0
	R	100
	B	100
Purple	G	0
	R	22.1
	B	100
White	G	100
	R	100
	B	100

kV) electrons, and the innermost phosphor (on the inside of the CRT face) is a green one that requires higher-energy (>10 kV) electrons to excite it. Thus, the red or green can be generated by the choice of beam voltage. Yellow to orange can be generated by simultaneous excitation of the red and green phosphors. The net result is a color CRT whose ruggedness matches that of a monochrome CRT and whose spectrum is limited to three colors. A disadvantage of the CRT is the requirement for rapid, precisely timed modulation of the beam voltage to produce displays with adequate resolution and color purity.

Beam indexing is a latent technology that promises to provide color capability in airborne CRTs. Like the beam penetration tube, this technology does not require a shadow mask. In a beam index tube the face is lined with fine vertical stripes of four phosphors: the standard blue, green, and red plus a fourth one that emits ultraviolet or electrons when excited by the scanning beam from a standard single-beam gun. This emission is detected within the tube envelope, and feedback is used to track the beam position so that the beam current can be modulated to produce the correct color. Proper choice of beam current, coupled with precise knowledge of the beam location, allows a complete range of colors, matching those of the shadow mask CRT, but somewhat brighter than it since the attenuating effects of the shadow mask are not present.

Beam indexing technology CRTs are very attractive because of their simplified design and ruggedness. It is generally agreed that they will not find widespread use in the cockpit until the development costs are underwritten by a much larger market, such as portable televisions.

Stroke and raster are the two methods of generating the image on a CRT. As is often the case, each has its advantages and disadvantages. Stroke generation of the image is an adaptation of the method used in oscilloscopes to generate the trace. That is, the electron beam is deflected to produce a symbol like a waveform on an oscilloscope. Stroke has several major advantages including sharpness and brightness. Improved brightness is achieved by using a slower writing speed (beam velocity across the CRT face) than is used in raster generation. Because of this relatively high brightness, stroke-generated images find extensive use in head-up and helmet-mounted display CRTs where the image must be viewed in the presence of a very bright background.

Raster-generated displays also have attractive features when compared to stroke-generated displays. Raster is well suited to displays requiring areas of continuous tone such as an electronic attitude display indicator (EADI) where one color represents the sky and another the ground. Raster is also well suited to overlapping display elements where a symbol must appear to be in front of the background and not blended in. A disadvantage of raster displays is the "jaggies" that occur whenever a line in an image crosses raster lines at a very shallow angle. This problem can be solved by dithering the line between adjacent vertical display elements coupled with careful modulation of the brightness and by the use of higher-resolution (more narrow line-to-line spacing) CRTs.

A solution to the stroke versus raster trade-off question is their simultaneous use, which allows the designer to use the best features of both methods. Typical of this approach are the displays in the B-757, B-767, A-310, and A-320. This hybrid approach has merit in many applications.

CRT displays quickly gained wide acceptance in part because the display designers solicited the expert opinion of experienced line pilots who subjected the CRT displays to extensive simulator and flight testing. Since the combination of CRTs and digital avionics is such a powerful one, there is a natural tendency by the display designer to show almost everything possible about the aircraft and its flight path. However, the initial evaluations of candidate CRT displays showed that a conservative design approach was the correct one. The designer should use his or her imagination to develop display formats that are clear and understandable and that present information, not just data. Remember, there is always the option of calling up another display to amplify and clarify a particular situation.

Luminous Flat Panels

Luminous flat panels are emerging as a major technology in cockpit displays. One reason for their popularity is given in the latter part of their name—flat panel. The volume of a flat panel display is only a fraction of that of a CRT display, and the requirement for clearance behind the instrument panel is much less than that of a CRT or, in many cases, even an equivalent electromechanical indicator. Reduced volume and rear panel clearance are significant advantages for the display designer since they allow greater flexibility in display device placement. Displays can be put where they are needed, not just where they can fit.

Another advantage of luminous panels is the absence of the high voltage required for CRTs and its associated problems. The potential for CRT catastrophic implosion is eliminated. Relative to CRTs, flat panels are very rugged, which eliminates the added complexity of shock mounts.

Flat panels have uniform resolution and brightness across the screen. There is no blurring near the edges of the image, and it is uniformly bright.

All of these features, singularly and collectively, are leading to an explosive growth in the application of flat panel displays. There is a broad consensus that, in the long term, flat panels will take over almost all of the functions now performed by CRTs. Many people view CRTs as only the first step in cockpit electronic displays while flat panels are the predicted destination.

Light emitting diodes (LEDs) were first used in the original digital watches where a button was pressed to display the time. Since then the technology has improved and is widely used today in airborne flat panel displays.

LED displays have a low quantum efficiency, so relatively high power is required to produce acceptable brightness, especially at high altitudes in full sunlight. This inefficiency also means that much of the input power is dissipated as waste heat and may be a factor in cooling the cockpit. Since LEDs are solid state devices, their reliability is excellent, much better than that of CRTs.

Flightworthy displays as large as 5 by 5 inches (7 in diagonally) have been built and demonstrated. A black layer is used to improve contrast in high ambient light conditions.

State-of-the-art displays have 64 elements per inch and can generate three colors: red, green, and yellow. Actually there are only red and green LEDs; yellow is produced by simultaneously exciting red and green. The lack of a full range of colors is a limitation of LEDs. However, the three available colors can satisfy many cockpit display requirements, particularly as status, caution and warning displays.

Work has begun on using LEDs in switch legends. The intrinsic color capability naturally suggests such uses: In an emergency push the red switch first. Also pay attention to yellow (caution) switches.

Electroluminescence is another emerging technology for flat panel luminous displays. A 6-in diagonal display has been constructed and driven at video rates. Resolution is 68 lines per inch. There are presently two limitations: brightness and limited color (there is no practical blue phosphor for flight applications). Brightness, of course, is the most severe of these limitations, and research is under way to enhance the brightness to a level satisfactory for operation in full sunlight. Limited color is less of a limitation; however, research on practical blue phosphors is under way. A full color thin film electroluminescent (EL) display suitable for use in the office or laboratory has been built and demonstrated; however, flightworthy EL displays are still not available. There are many ground applications for the existing limited-color-capability EL displays, but if these displays are ever to be a serious candidate for replacing CRTs in the cockpit, full color is essential.

Gas discharge displays are the last example of the flat panel luminous displays to be examined. The phenomena of gas discharge has been known for decades and extensively exploited in ground-based applications where alternating current (ac) voltage was used for excitation. For airborne applications, direct current (dc) voltage is preferred since it can achieve higher brightness. Even so, efficiency is relatively low, about 0.5 lumen per watt (lm/W). Resolution is about 40 lines per inch.

Gas discharge displays have found only limited application in cockpits. Perhaps the best example is the L-1011-500 flight management system control/display unit. Brightness of this display is 1000 candela per square meter (cd/m^2).

Gas discharge displays require a current-limiting resistor which, when the display is operating at high brightness levels, dissipates significant heat. The aircraft designer must include provisions for handling this heat dissipation load.

A Look at Liquid Crystal Displays

Liquid crystal displays (LCDs) are the third and final type of electronic display to be examined. (There are, of course, other types of electronic displays, but they are either unproven or not used enough to warrant mention.)

The liquid crystal phenomenon comes from altering the anisotrophic structure of long organic crystals by applying electrical fields so that crystals of these materials will undergo change in their optical, magnetic, and electrical characteristics. In the case of optical properties, when an electrical field is applied, the crystals change from transmissive to reflective or vice versa depending on the type of material (Fig. 3.2).[2]

LCDs share many attractive characteristics with EL displays. Both types of displays are flat panels, solid state, compatible with digital drive circuits, can be matrix addressed, and suffer no loss of image sharpness at their edges. LCDs, though, have an additional attribute of low power consumption. The first-generation LCDs are passive; that is, they only reflect, transmit, or absorb the ambient light. Consequently, there is no problem with heat dissipation, which is a real bonus in the cockpit, an area of the aircraft that is chronically plagued by high (by crew standards) temperatures. Because LCDs are passive, there must be a light source available when the cockpit is dark. Frequently small lights, whose brightness is controlled by a potentiometer, are in the bezel of the LCD.

The second-generation backlighted, color LCD is beginning to be applied in aircraft now being developed. The resolution and brightness of these second-generation LCDs rival that of the CRTs while offering lighter weight, reduced depth behind the panel, and lower power. Ad-

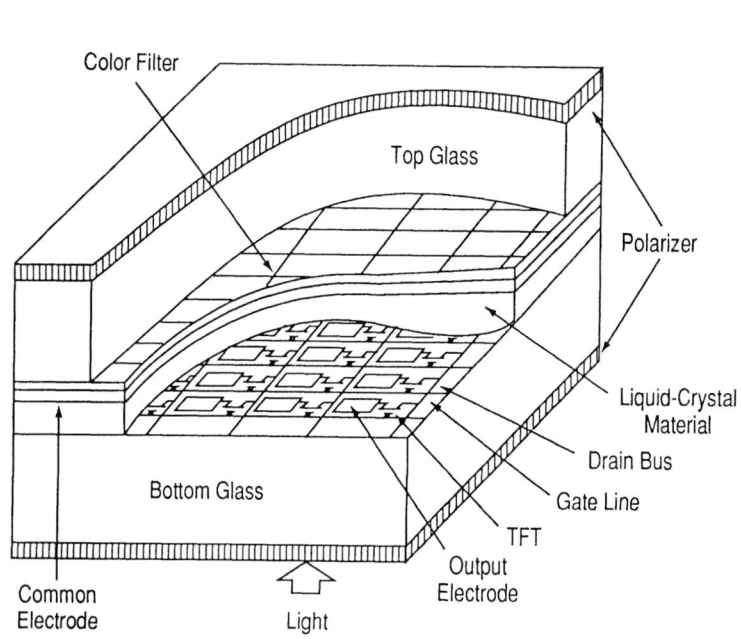

Figure 3.2 Cutaway view of an active matrix LCD display. (Reprinted from *Electronics*, May 1989, copyrighted 1989 by Penton Publishing, subsidiary of Pittway Corporation.)

ditionally, the reliability of LCDs is generally substantially better than that of CRTs. Figure 3.2 shows the components of a typical color LCD. The display surface is divided into rows of picture elements (pixels). For state-of-the-art avionics applications there are typically 1296 rows of 1296 pixels each. For each pixel there is a filter of one of the three primary colors: red, green, or blue. Between that filter and the backlight is a shutter of liquid crystal material. If the pixel is to be lighted, the shutter is transparent; if the pixel is to be dark, the shutter remains in its normally opaque condition. The transparency/opacity of the shutter material in each pixel is controlled by the voltage applied by the display drive circuit to that pixel. Most backlighted LCDs have redundant light sources to increase reliability.[2]

One troublesome trait of LCDs is poor off-axis viewing. Contrast and sharpness fall off rapidly as the viewer moves farther from the normal view of the display, although LCDs with twisted axis crystals promise to substantially remedy these shortcomings.

Another member of the LCD family is dichroic displays. These displays are generally the inverse of conventional LCDs in that they have white or bright color patterns on a dark background. Dichroic displays are switched on and off like all the other LCD displays. In the on state, the liquid crystal molecules are aligned so that the dye in the crystal will reflect the ambient light. In the off state, the molecules are unaligned and the ambient light is transmitted through the crystal to be absorbed by the dark background. To date, these displays have not been matrix addressed and, therefore, are limited to the familiar alphanumeric displays.

Comparing the Display Media

The previous sections in this chapter have reviewed three basic types of electronic displays: CRTs, LEDs, ELs, and LCDs. In this section, these displays will be compared with each other for a number of important performance parameters to aid the designer in selecting the optimum display media for a given application. Table 3.2 summarizes the comparisons.

Average luminance, or brightness, is a major consideration in selecting a cockpit display. Full sunlight is deceivingly bright, and many candidate displays that appear sharp and full of contrast in the laboratory completely wash out in a sunlit cockpit. A quick scan of Tables 3.2 and 3.3 shows that sunlight is orders of magnitude brighter than any of the displays. This gross mismatch in brightness is accommodated through the use of narrow bandpass filters which exactly match the wavelength of the light radiated from the display, thereby blocking out all other wavelengths of sunlight. Of course, brightness is not an issue with reflective LCDs.

Resolution is another factor in any display, but all of those reviewed here have adequate and approximately equal resolution, so it is not generally a trade-off item.

The issue of color is always a hotly debated one. There is generally

TABLE 3.2 Comparison of Major Features of Display Media

	CRT	LED	EL	Backlighted LCD
Avg. luminance (fc)	100*	5	27	†
Resolution (lines/in)	200	100	68	200
Color	Full	Red, green, yellow	Red, green, yellow	Full
Viewing angle	±80°	±80°	±80°	±50°
Power (W/in^2)	5	8	0.6	2×10^{-6}‡
MTBF (h)	3000–15,000	16,000	10,000§	10,000 +
Temp. range (°C)	−20–+70	−20–+70	−30–+100	−15–+40

*Assumes a 1/525 duty cycle (525 lines/frame).
†Luminance determined by backlighting intensity.
‡Does not include backlighting power.
§Time to half brightness.
SOURCE: ADAPTED FROM ROBERTSON, JAMES B., "FLAT-PANEL DISPLAY TECHNOLOGY," TUTORIAL PRESENTED AT THE IEEE/AIAA 7TH DIGITAL AVIONICS SYSTEMS CONFERENCE, 1986.

TABLE 3.3 Sample Values of Luminous Intensity and Luminance

Luminous intensity (cd/m^2)	Luminance (lm/m^2)
Sun surface: 2,000,000,000	Outside, clear day: 100,000
Incandescent lamp: 10,000,000	Outside, overcast: 10,000
Fluorescent lamp: 6,000	Office: 100
Moon: 2,900	Moonlight: 0.2
	1 lm/m^2 = 1 lux (lm/m^2)/10.764 = lm/ft^2
	1 lm/ft^2 = 1 fc
	1 lm/ft^2 = 1 ft · L (for a perfectly reflecting surface)

no question about the value of color in a display; there is only the question about whether it can be achieved at a reasonable technical and economic cost. Only CRTs and LCDs presently offer full color capability. LEDs can produce only three colors: red, green, or yellow (through the combining of red and green). EL displays come only in red, green, or yellow, depending on the choice of phosphors and the drive voltage.

In color electronic displays, careful thought must be given to what happens when one or more colors are lost. The loss of color is clearly a certification concern in civil aircraft. It must be demonstrated to the certifying authorities that, with one or more colors failed, adequate information can still be conveyed to the crew through the careful choice of symbology and primary colors used to generate the symbols. In fact, FAA Advisory Circular 25-11 requires that in the case of the loss of one or more colors the display does not have to be useful but must not present misleading information.[3]

The gray scale levels in all of the candidate media are adequate to provide continuous shading in an image without any evidence of banding (discernible boundaries between adjacent areas of consecutive discrete brightness levels).

Power, or more specifically, power dissipation, is always a concern in the cockpit area; however, none of the displays discussed here generally dissipate enough power into the cockpit to cause any difficulty. CRT displays must be cooled; however, the cooling air is not exhausted to the cockpit. If power consumption is a concern, either ELs or passive LCDs would be a logical choice.

All of the displays show excellent reliability, equal to or better than that of electromechanical indicators.

All of the candidate media, except LCDs, have an adequate operating temperature range. LCDs exhibit sluggish response at low temperatures ($< -18°C$, or $0°F$), so auxiliary heating is recommended for aircraft applications. Conversely, LCDs have reduced contrast at high temperatures, so cooling must also be provided; however, available cockpit cooling should be adequate.

One very important consideration in selecting displays is the depth required behind the instrument panel. Generally, the only time this is a problem is in the case of CRTs, which require up to 19 in behind the panel. Such deep displays sometimes limit the locations in which they can be installed. The relatively shallow depth requirements of all of the other candidate media allow almost complete freedom in positioning them on the instrument panel.

In addition to the factors listed in Table 3.2, there are several other points that need to be addressed in making display media trade-offs. Matrix addressed displays, which are all of the ones listed in the table except CRTs, have a graceful failure mode in that the loss of one element, line, or column is not catastrophic and would not generally result in a substantial loss of data. Matrix addressed displays are intrinsically digitally compatible. However, the problems of interfacing a CRT with a digital circuit have long been solved, so this feature of matrix addressed displays is no substantial advantage.

Another intrinsic feature of solid state displays is ruggedness. While CRTs must be rugged to operate properly in an aircraft environment, the other display media in the table can be readily installed in an aircraft without special design or mounting provisions.

Finally, cost is frequently not an issue at all. The cost of the display element itself is relatively small, so it has only a marginal influence on the total cost of a display system. The technical merits of display media are usually the determining factors in making a selection.

Multifunction Keyboards Save Space, Offer Flexibility

As the cockpits of modern aircraft have more controls jammed into them, the point is reached where there is no more space. Multifunction keyboards (MFKs) offer a very attractive solution to this space problem wherein a single panel of switches performs a variety of functions depending on the phase of the mission or the keyboard menu selected. Since MFKs are based on display technology discussed earlier in this chapter, their design uses the same principles and, in some cases, it may be possible for a display to double as a tactile input device.

Multifunction keyboards can be implemented in several ways. The first two ways use LEDs or LCDs in panels in a central location. Designs using LEDs have arrays (typically ranging from five rows of three switches to seven rows of five switches) of standard-sized pushbutton switches with legends built into the surface of the switches. These legends are generated by small arrays of matrix addressable LEDs in the switch cover. Color can be used if desired. The legends are changed by the user or automatically by the aircraft as the mission phase changes. Once the top-level function of the keyboard has been established for a given mission phase or menu, the legends may change as the keyboard is subsequently used in secondary functions or submenus.

For MFKs designed using LCDs, the panel is usually a continuous smooth surface with the individual switch function and area established by the matrix addressed LCDs. Since LCDs have a graphics capability, it is also possible, for example, to enter a numerical value by pressing the surface of an LCD panel in the area where a portion of a curve is displayed rather than entering a numerical value by using a keyboard. As is the case for LCD displays, some provision must be made for lighting the panel when the cockpit is dark.

CRT displays also have the capability to serve as MFKs. Like the solid state display techniques above, there are two ways of accomplishing this. The most common way is for the CRT face to contain function labels correlated with adjacent switches mounted in the bezel surrounding the CRT. Of course, it is very easy through software to change the switch function and the associated label on the CRT.

The second method of implementing MFKs on CRTs is to overlay the CRT face with a touch-sensitive screen. Once again, the CRT face contains various switch function legends, and the corresponding function is selected by touching that portion of the screen over the switch label. A variation of this approach is to replace the touch-sensitive screen by an array of infrared diodes and detectors on opposite sides of the CRT so that when the CRT face is touched, the infrared beam is interrupted, thereby creating a signal that is functionally equivalent to pressing a switch.

A major consideration in any tactile input device is that of ensuring that the proper characters or data are entered. The most common way of ensuring this is to display the input on a scratch pad (a small portion of an adjacent or nearby, easily viewed display) as it is entered and then requiring a separate Enter command after the correctness of the input is verified. Such provisions are especially necessary where tactile input devices will be operated in turbulent or unsteady flight conditions. To further improve the correctness of tactile inputs, the entry should occur upon lifting of the finger as opposed to when the surface is initially touched (or the infrared beams interrupted). This technique allows the operator to fine tune and/or steady the position of his or her finger before making the selection.

Head-Up Displays Offer Innovative Options

A head-up display (HUD) is another device made possible by electronic display media. Using a high-brightness CRT and a special type of mirror called a combiner, it is possible to build a display that projects some of the information normally on the primary flight display and selected systems or weapons data into the line of sight of the pilot without substantially dimming or obscuring the outside view. A HUD allows the pilot to simultaneously see critical aircraft information while viewing the outside scene directly ahead.

The basic elements of a HUD are shown in Fig. 3.3a. Every HUD contains, as a minimum, a display generator and a combiner. In current HUDs the display generator is a high-intensity CRT with a P43 (green) phosphor. (Research is under way on alternative generators such as a matrix addressable LCD array illuminated by a high-intensity lamp.) The combiner is a mirror with several unusual properties: The reflective coating is highly wavelength selective, centered on the wavelength of the emission from the CRT phosphor and clear to other wavelengths, and the coating is also highly selective in angle of incidence so that only that light impinging within a very narrow range of angles will be reflected. The combiner is sometimes incorrectly referred to as a *hologram,* but it contains no image information as found in a true hologram. Combiners are constructed photographically by exposing a film of dichromated gelatin to crossed laser beams and then developing the resulting diffraction pattern image. Since the gelatin is organic, is must be protected from the environment, especially from high temperatures.

High-performance aircraft HUDs use one of two basic designs for the combiner. The simplest is the single element combiner shown in Fig. 3.3a. The actual gelatinous combiner element is the middle layer of the three-layer combiner assembly. This single-element version is somewhat limited when compared to the three-element combiner discussed below. Transmission of the outside scene is higher in a single-element combiner than in a three-element combiner. Transport air-

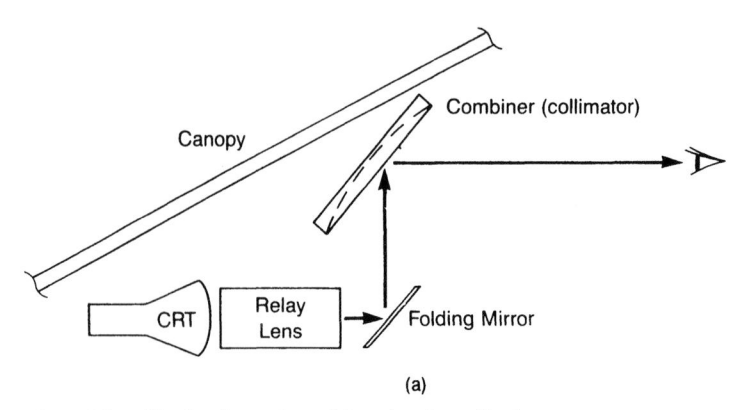

(a)

Figure 3.3a Single element combiner head-up display.

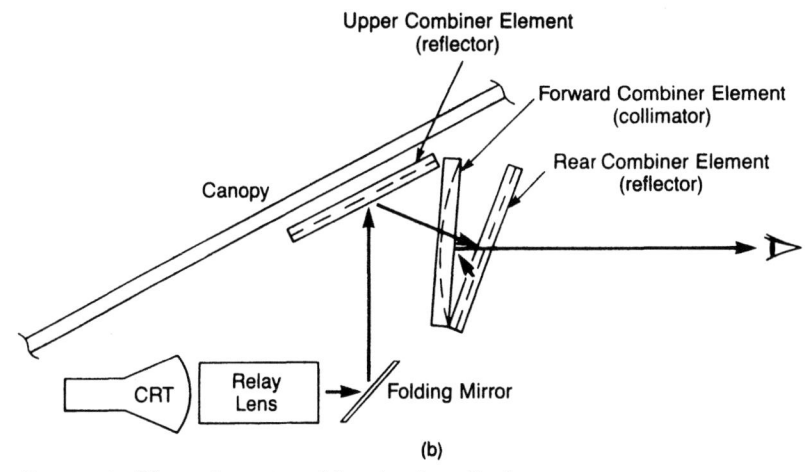

Figure 3.3b Three-element combiner head-up display.

craft use the single-element design since it is particularly amenable to installation in a transport cockpit.

A three-element combiner design, shown in Fig. 3.3b, is used on some high-performance aircraft to achieve better producibility, simpler overall HUD design, freedom from secondary reflections, and a large field of view. This design has achieved a 30° horizontal and 20° vertical field of view. All three elements in this design contain gelatinous combiners as the middle layer, but only the forward element is curved to collimate the image from the CRT.

An extremely important consideration in HUDs is the information content and format of the projected display. Only essential information must be shown, and it must be in an instantly recognizable form.

Figure 3.4 shows a typical commercial transport HUD display during landing. Note the relatively uncluttered (to the eye of a trained, expert pilot) display. The symbology is carefully chosen to present the maximum amount of information with the minimum amount of symbology and consequent obscuration of the outside scene.

A HUD is monochromatic for maximum brightness. Color HUDs are controversial for two reasons. First, there may be some loss of brightness, although brightness is becoming less of an issue as color CRTs improve. Second, colors may be confused with or lost in the natural exterior scene. Only extensive flight testing and field experience will ultimately resolve these questions.

Perhaps the major practical issues with HUDs are their large volume and the necessity to be mounted in the cockpit so that the combiner is in the line of sight of the pilot. On high-performance aircraft, the HUD is mounted at the top of and behind the instrument panel so

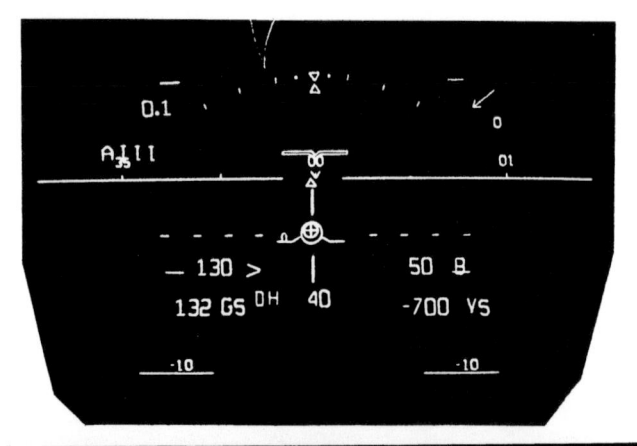

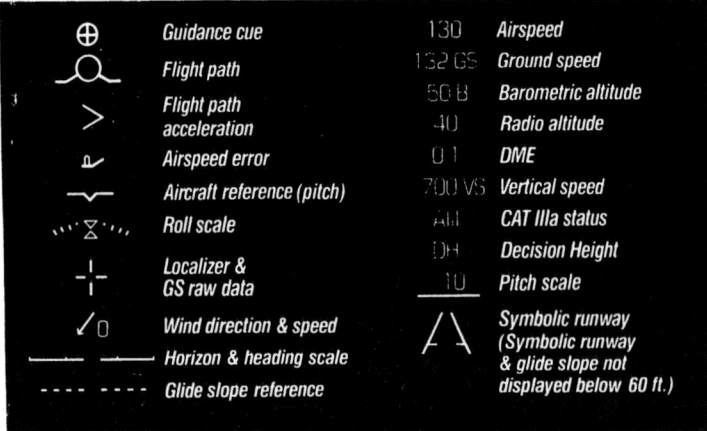

Figure 3.4 Typical commercial transport HUD during approach. (*Courtesy of Flight Dynamics, Inc.*)

that the combiner is between the top of the panel and the canopy in the pilot's line of sight when looking straight ahead. For civil transport aircraft, a HUD is mounted above the seat of each cockpit crew member, and the combiner is hinged to swing down into the line of sight when the HUD is in use, generally only during approach and landing. An alternative design for civil transports is a single-element combiner that pops up from the glare shield when needed.

Head Level Displays Offer a New Option

Head level displays (HLDs) are another option for cockpit displays. An HLD avoids the physiological limitation on eye refocusing time (as high as 200 ms) by placing directly below the HUD or top edge of the instrument panel a display in which an image and supplemental alphanumeric information are focused at a long distance, say about 50 m. Thus, the need for the pilot to refocus his or her eyes to scan at least some information inside the cockpit is eliminated. Typically the HLD will contain a radar or infrared image of the outside scene or, perhaps, a digital map. An HLD uses a high-intensity lamp coupled with dichroic filters to sort the white light into red, green, and blue and with optics to collimate and fold the light. The red and green bands are each modulated by liquid crystal shutters in which each pixel is either opaque or transparent as required to generate a color image. An excellent description of a flightworthy HLD can be found in reference 4.

Helmet-Mounted Displays

The high brightness, excellent resolution, and capability to be built in lightweight, small sizes has led to the development of helmet-mounted displays (HMDs) using monochrome CRTs. HMDs have a major advantage over HUDs since critical aircraft and stores information is in the pilot's line of sight at all times, not just when he or she is looking straight ahead. HMDs are especially attractive for use in high-performance aircraft where the pilot is frequently engaged in low-level penetrations and air-to-air or air-to-ground missions and must pay almost complete attention to the outside situation.

HMD display formats are very similar to those of HUDs except for the addition of helmet-pointing azimuth and elevation information and vectors showing where the last target of interest was prior to looking down into the cockpit or searching for another target. A typical HMD display is shown in Fig. 3.5.

Designing an HMD requires careful consideration of two additional important factors: weight and helmet aerodynamics. During high vertical acceleration maneuvers such as tight turns and ejection from the aircraft (on the order of 9 g), the helmet can become a very heavy object, indeed, which leads to the mandate to design absolutely minimum-weight helmet-mounted optics. Some advanced helmets have been built using lightweight materials for the basic helmet shell and for the optics that have resulted in a helmet with HMD that weighs less than many standard helmets without an HMD.

Immediately following ejection, the helmet is exposed to a high-speed airflow which can generate substantial lift when flowing over a properly shaped object. Thus, the designer must ensure that the helmet is *poorly* designed from an aerodynamics perspective to ensure that it does not generate any lift and thereby suddenly pull on the pilot's neck immediately following ejection.

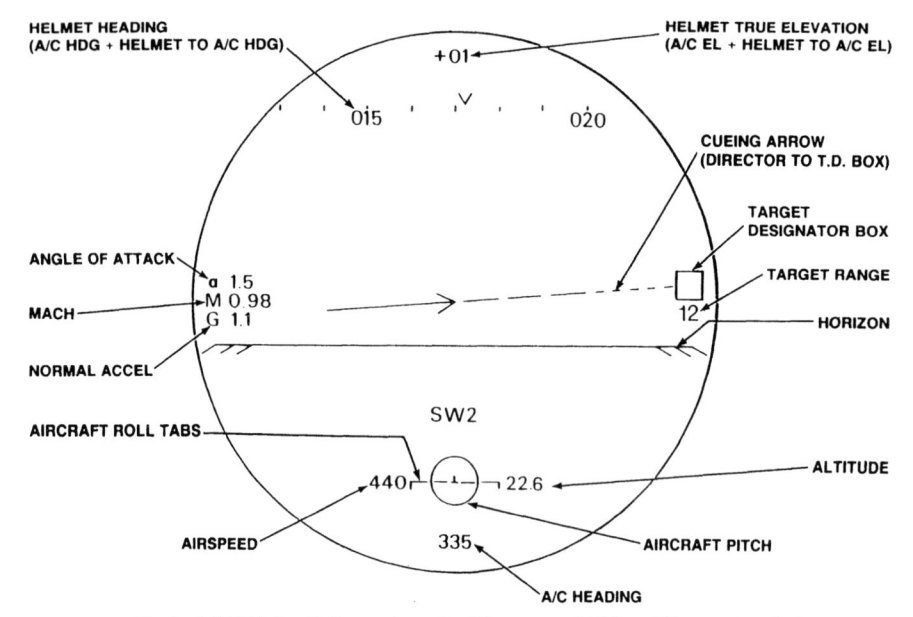

Figure 3.5 Typical HMD for fighter aircraft. (*Courtesy of Kaiser Electronics Co.*)

Another complicating element in HMDs is the need for a helmet tracker that can determine the direction the pilot is looking. Pilot look angles are used to steer aircraft auxiliary imaging systems such as radar and infrared and/or to designate a target. Most helmet trackers use three mutually orthogonal magnetic fields, each with a dedicated sensor, to determine the direction in which the helmet is pointing. Because the sensors are magnetic, nearby metallic structures can affect the performance of the tracker. Thus, it is necessary to calibrate the tracker in every aircraft in which it is installed.

HMDs have also been examined as a means of permitting maintenance personnel to have both hands free and eyes fixed on the repair task at hand while simultaneously viewing maintenance drawings and procedures.

Night Vision Goggles

For rotorcraft and attack aircraft, night vision goggles provide a daylight-equivalent view of the outside scene. The cornerstone of night vision goggles is an image intensifier tube which uses a phosphor to convert the incoming visible or near-infrared photons from the outside scene to electrons that are multiplied by channel electron multipliers. The multiple electrons then strike a green phosphor to generate a bright, monochromatic rendition of the external scene. Detailed requirements for night vision imaging system (NVIS) requirements are documented in MIL-L-85762A.[5] The specification establishes two types of NVISs. Type I uses only a binocular goggle with a phosphor screen image. All of the pilot's vision is through the goggle. Type I is used generally in rotorcraft. Type II NVIS have the same goggle as Type I, but a provision is made to allow the pilot to directly view the instruments through a combiner positioned below the goggle. Type II is commonly used on fixed wing aircraft.

Because of the enhanced sensitivity of NVIS to red and near-infrared light, there are restrictions on the type of cockpit lighting that can be safely used to avoid washing out the outside scene. According to MIL-L-85762A, rotorcraft cockpit lighting is limited to blue, green, and yellow. No orange or red is allowed. The filter used in rotorcraft NVISs (Class A) sharply attenuates (>99 percent) of all incoming radiation at wavelengths less than 595 nanometers (nm). Since red wavelengths are not attenuated, any red cockpit lighting would wash out the NVIS and render it useless. Fixed wing aircraft NVISs (Class B) can accommodate the full range of cockpit lighting provided that red light originating in the cockpit does not have a wavelength greater than 625 nm where the NVIS filter begins transmitting.

Laying Out the Cockpit

Laying out the cockpit is one of the most difficult jobs faced by the avionics designer. There are three factors which contribute to this difficulty: Space in the cockpit and on the instrument panel is extremely limited (one cockpit expert has described the instrument panel as the most valuable real estate in the world),[6] the cockpit is often the hottest spot in the aircraft, and the layout of the instruments is driven by equal parts of engineering and human factors. Failure to recognize human factors in laying out the cockpit guarantees the crew will be limited in their ability to interact with the aircraft and its systems.

Sound cockpit design continues the top-down process begun in Chap. 1 but shifts the focus to a more specialized level for defining the requirements for the cockpit and the displays. The processes essentially continue the top-down, decompositional approach described in Chap. 1 but under the guises of terms such as crew workload analysis and timeline analysis. Typical cockpit design techniques include action/information requirements, timelines, and workload analysis.[7]

The definition of action/information requirements builds on the results of the top-level system design processes (e.g., mission profiles and mission scenarios) and adds information to that found in functional flow diagrams or decision/action diagrams. This process is rel-

TABLE 3.4 Excerpt from a Typical Action/Information Requirements Worksheet[7]

Functional requirement	Action requirement	Information requirement
1.0 Initiate preapproach procedure	1.0.1 Review approach information	1.0.1.1 Approach orientation 1.0.1.2 Approach constraints Requirements Obstacles/hazards Weather Minima
	1.0.2 Coordinate with approach control	1.0.2.1 Communication Path designation Unique limits/constraints Environmental conditions Barometric pressure

atively informal, qualitative, and unstructured. Table 3.4 shows an excerpt from a typical action/information requirements worksheet.[7]

Timelines, a chronological depiction of all tasks performed by the crew, are a widely used formal, quantitative technique for designing cockpits. Timelines are traceable to the functional requirements and, therefore, aid in ensuring that all tasks and their details are identified. In the case of multiple crew members, separate timelines are prepared for each of them. The time estimates used in laying out a timeline are usually based on previous similar tasks performed in similar environments. Otherwise, it is necessary to resort to fundamental task time studies. (It is important for these task time studies to be conducted in high-fidelity simulators to ensure the results are fully applicable to the cockpit being designed.)

Workload analysis is based on the integrated, accumulated timelines. The focus is on individual crew members to identify the workload on each perceptual and motor channel. The channels generally analyzed in a typical workload analysis include external vision, internal vision, left hand, right hand, feet, cognition, audition, and verbal. The resolution of the timelines will change depending on the level of activity under analysis. In very busy times the resolution should be 1 s or less, while at less busy times 5-s resolution should be satisfactory. The precise summation effects of simultaneous activity in several channels is difficult to assess and requires the intuitive skills of a human factors specialist. If a more detailed workload analysis is required, it may be necessary to conduct simulator experiments. Workloads of 100 percent are not acceptable, even briefly, and workloads of 75 to 100 percent should be avoided. Of course, the reverse of these high-workload situations, crew boredom, can also be a problem at times.

The analyses described above are the foundation for laying out the cockpit. A basic tenet for this very important process is to make the cockpit user friendly so that operation of the avionics and the aircraft are easy and foolproof. The designer should strive for a quiet, dark cockpit—one that signals the crew only when action or input is required or a system status has changed. In order to reduce the number of controls and associated cockpit space, the designer is encouraged to use multifunction controls and displays wherever possible.

In laying out the cockpit, the two principal guiding documents are MIL-STD-203 Aircrew Station Controls and Displays: Assignment, Location, and Actuation of, for Fixed Wing Aircraft, and FAR 25.1303 Flight and Navigation Instruments. As noted earlier, cockpit design is strongly influenced by human factor considerations, so these documents contain common sense—things that are obvious to most designers but have been written down anyway to be sure they are not overlooked and are uniformly applied from one aircraft design to the next.

MIL-STD-203 calls for consistent operation and uniform actuation for all controls in the cockpit. Forward, upward, or clockwise operation of controls should increase performance and, conversely, backward, downward, or counterclockwise operation should decrease performance. Where switches are mounted on panels having slopes within 30° of vertical, they must be flipped upward to enable a function. Where switches are mounted on panels with slopes greater than 30° from the vertical, they must be flipped forward to enable a function. Switches must be capable of being operated with gloves on.[8]

Displays should be perpendicular to the line of sight of the operator, wherever possible, and never more than 45° from the line of sight when looking straight ahead. Displays and controls for like functions should be grouped and should encourage viewing and operation from left to right, top to bottom, or back to front (closest to the operator).[8]

Federal aviation regulations dictate certain design constraints on civil aircraft cockpits. For example, FAR 25.1303 Flight and Navigation Instruments requires each pilot station to be equipped with an airspeed indicator, an altimeter, a rate-of-climb indicator, a gyroscopic rate-of-turn indicator, a gyroscopically stabilized bank and pitch indicator, and a gyroscopically stabilized direction indicator. Additionally, instruments visible from each pilot station must include a free air temperature indicator, a clock, and a magnetic compass direction indicator.[9] FAR 25.1321 Instruments: Installation, Arrangement and Visibility prescribes the arrangement of the most important of these instruments into the familiar T. Paragraph 25.1321 (b) reads in part:[10]

(1) The instrument that most effectively indicates attitude must be on the [instrument] panel in the top center position;

(2) The instrument that most effectively indicates airspeed must be adjacent to and directly left of the instrument in the top center position;

(3) The instrument that most effectively indicates altitude must be adjacent to and directly right of the instrument in the top center position and

(4) The instrument that most effectively indicates direction of flight must be adjacent to and directly below the instrument in the top center position.

FAR 25.1322 Instruments: Installation, Warning, Caution and Advisory Lights requires the following color code for indicator lights: red for warning lights indicating a hazard that may require immediate corrective action, amber for caution lights indicating a condition that may require future corrective action, and green for safe operation.[11]

As stated earlier, human factors play a large part in the design of a cockpit. An outstanding example of a cockpit that has been designed with careful attention paid to human factors is on the Boeing B-757 and B-767 aircraft. The primary flight displays (the basic T described above) and the system status monitors use color CRTs. Display formats can be selected by the crew from a menu of possible options. When all systems are normal, the cockpit is quiet and dark. Lights are illuminated or aural warnings are given only when action is required from the crew. Preflight systems checkout is accomplished by pushing each lighted switch in the overhead panel. The correct preflight sequence of switch operation is ensured by going down (forward) each column of switches and following a left-to-right pattern of column scan. When all lighted switches are turned off, preflight checkout has been successfully completed. Another important feature of the B-757 and B-767 cockpits is commonality. Both aircraft have the same cockpit forward of the front bulkhead.

The use of CRTs and automation in the B-757 cockpit generally has been well received by the crews. A comprehensive, long-term study of 200 B-757 pilots from two United States airlines revealed wide acceptance of the cockpit and only a few concerns, primarily about the workload associated with entering information into a multifunction control/display unit (MCDU), especially when very busy at low altitudes, and the loss of flying skills resulting from over reliance on the automation.[12]

A recent U.S. Air Force study of military aircraft has highlighted the importance of sound, logical, human-centered cockpit design. Based on the findings of the study, some practical design tips have emerged:

- Process the aircraft sensor signals so that the crew is presented information, not disparate data that must be synthesized and interpreted by the crew. Where possible and appropriate, use pictures, icons, and diagrams instead of words. Have available to the crews, through call-up menus if necessary, complete information on aircraft, systems, mission, and stores status.

- Electronic-display-based menus should have a Home button and a button to return to the previous (higher-level) page. Items frequently changed via electronic menus (radio and navigation aid frequencies, altimeter setting, etc.) should be on the first (or certainly no more than the second) page of a menu. Electronic menus are well suited for making selections among four or more choices. For two or three choices, a toggle switch is satisfactory. Avoid similar names for dissimilar functions.

- Avoid cluttered displays. Especially avoid cluttered head-up displays. Label all parameters on a head-up display and keep related information together. (One USAF aircraft had related information farther apart on the HUD than in the cockpit.)[13]

Another important factor in designing the cockpit is to ensure continuity in the design to the maximum extent possible from one model, version, or derivative of the aircraft to the next. Careful attention must be paid to consistency and continuity with the previous or similar aircraft to achieve lower training costs and safer operation through reduced crew errors.

Voice Interactive Systems Are Another Way to Communicate

Voice interactive systems offer new opportunities for interface between the crew and the aircraft, especially in high-workload situations in single crew member aircraft. These systems fall into two fundamental categories: recognition and synthesis. Speech recognition has received the most attention in recent years as the newest of the two technologies, while synthesis is a mature and proven performer.

Both recognition and synthesis require extraordinary attention to human factors considerations relative to other cockpit technologies.

Speech recognition systems are capable of performing many functions in the cockpit; however, the functions should be selected only after very careful evaluation. Experience has shown that voice control is best applied to noncritical tasks such as requesting system status, tuning radios, and requesting maps to be displayed on a CRT. Clearly, voice recognition technology is not yet mature enough to handle urgent inputs or critical tasks such as firing weapons.

In applying voice recognition there are a number of unique problems not encountered in other cockpit controls. To begin with, each user of the system must generate a template of each word in the system vocabulary. The number of words in the vocabulary should be limited (generally 100 or fewer words, although this number is ever increasing as improved systems are being built) since the template for each word must be scanned every time a word is spoken (unless syntactical rules are used to limit the range of possible words to follow a given one), and generating templates can be time consuming. Generating a high-quality template requires that a word be spoken typically three to five times in the environment in which the word will be spoken in flight, including aircraft background noises and wearing an oxygen mask. It is also important that the microphone have the same electrical characteristics as the flight microphone and that it be positioned in the same location relative to the pilot's lips during template generation as it will be in flight. Duplicating the acceleration environment is desirable but not usually practicable. There is not yet a way to artificially stimulate the stress encountered in combat or its influence on pronunciation. Thus, stress remains one of the major causes of incorrect recognition of words. The templates are loaded into the aircraft voice recognition system prior to flight using a cassette tape or other portable memory medium.

It is possible to construct a speech recognition system that is speaker independent, vis-à-vis the speaker-dependent systems just discussed. However, the vocabulary is very limited, the templates require large amounts of memory, and signal processing is somewhat slower.

The most important figure of merit in a speech recognition system is the fraction of correct word recognitions. When a word is spoken, there are three possible responses by the system: correct recognition, confusion with a rhyming word, and rejection. Most systems can achieve correct recognition rates in the mid-90s percent in the case of a pilot experienced in interacting with a speech recognition system in a benign flight environment. As acceleration increases, the recognition rate decreases to less than 80 percent for a 9-g load on the pilot.[14] Another important consideration in using speech recognition systems is that the words must be spoken individually, with a pause between successive words. Continuous speech recognition systems are on the horizon, but none have been examined for use in a cockpit environment. As a means of accommodating the less than perfect recognition capability, each word is either repeated aurally by a speech synthesis system or displayed in typed form on a CRT. If the word has been correctly interpreted, the pilot can enable its execution by either another simple spoken word like *Go* or by operating a switch. If the word has been incorrectly interpreted, it is repeated.

The response time and the recognition problems can both be reduced through the use of syntactical rules and logic. Figure 3.6 shows a tree structure developed using syntactical rules to speed up the performance of a voice recognition system by limiting the number of word templates that must be searched to find a possible match to a spoken word. Node 1 is the master node and contains all words that are permitted to be the first word in a spoken sequence. Based on the initial word and the syntactical rules, the second word must be one of those in the sets of words in the secondary nodes (e.g., nodes 2 to 4 and 6 to 13 in Fig. 3.6). For example, *Comm 1* refers to a radio, so logic would dictate that the words which follow are numbers representing the frequency to which the radio should be tuned. Similarly, it is illogical to expect the command *Gear down* at 30,000 ft. Syntactically, *climb* and *descend* are generally followed by *to,* and *turn* is generally followed by *right* or *left.*[14]

Flight tests in the USAF Advanced Fighter Technology Integration (AFTI) F-16 have provided early benchmark results on the performance of voice interactive systems. The systems tested on the AFTI F-16 routinely achieved 95 percent correct word recognition; however, this percentage was reduced under high g conditions to less than 80 percent. There was an "overall reduction in pilot workload when voice was employed."[13] Several of the caveats noted earlier were confirmed. Voice control is not suitable for time-critical tasks. The aircraft audio system must be high quality with low EMI and stable electrical characteristics. The impedance of the system used to generate the template and the real aircraft system should closely match.

Voice synthesis systems have been used in cockpits for many years. The most familiar application is the "Pull up, pull up" message announced by the ground proximity warning system. As is the case for voice recognition systems, the application of voice synthesis systems must be carefully considered to avoid diminishing their effectiveness. Since the preferred use of voice synthesis is during high-workload periods, messages should be brief and in a telegraphic format. Messages should be restricted to conditions that (1) require immediate correc-

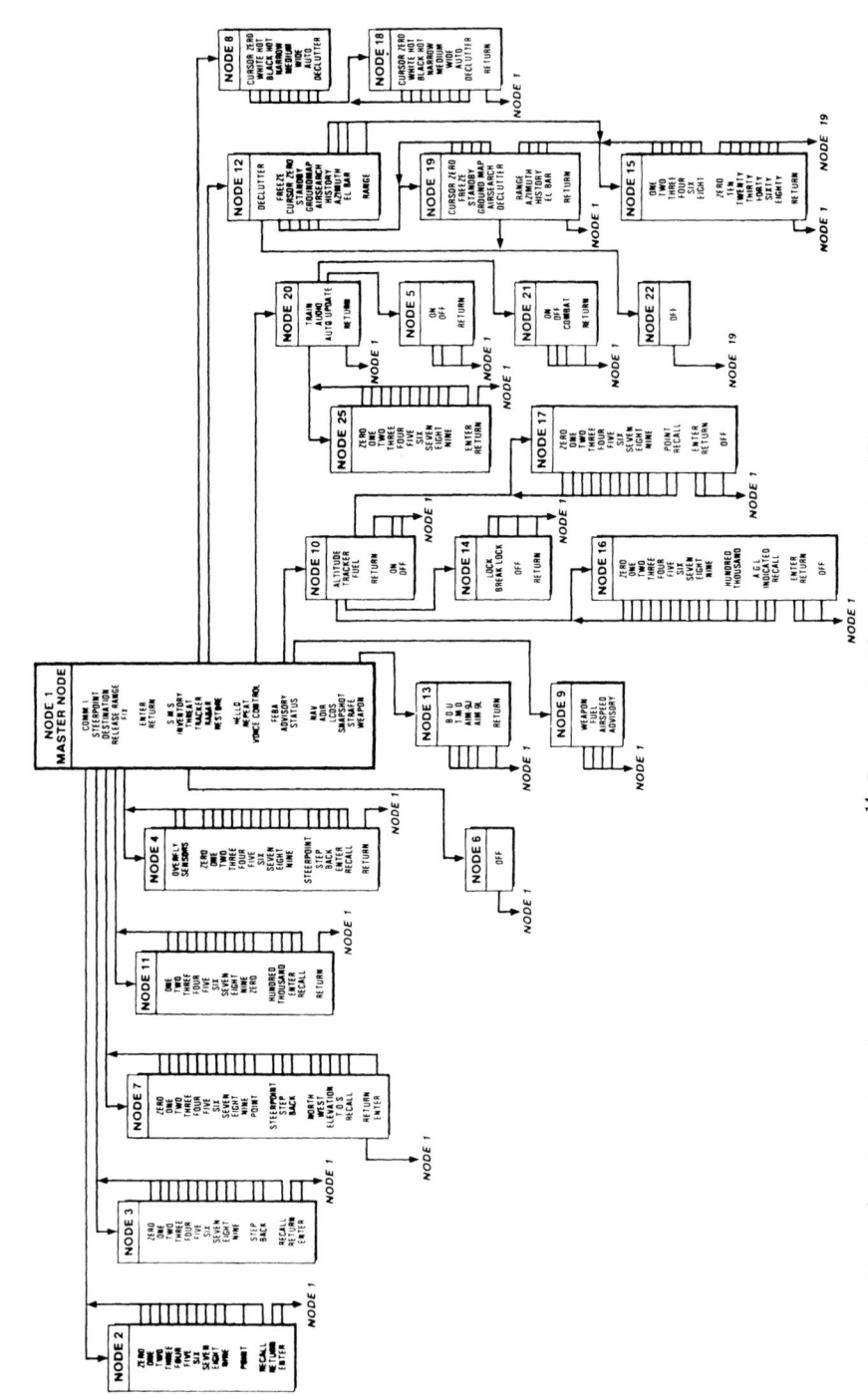

Figure 3.6 Example voice interactive system syntax tree.[14] (Reprinted with permission © 1987 from SAE paper 871751. Society of Automotive Engineers, Inc.)

tive action or have an immediate impact on the safety of flight or (2) will soon reach this condition if not corrected by the pilot. These messages should be preceded by a warning tone and have a concurrent visual display. There is a continuing debate without consensus on the gender of the warning voice.

Voice synthesis systems must include logic in the central processor that will resolve conflicts between nearly simultaneously received messages to be converted into synthetic voice. This resolution is generally achieved by a first in, first out protocol.

From Esoteric to Exoteric: Display Terms

Display technology brings with it a collection of esoteric terms that are confusing to the uninitiated. In this section, some of the basic concepts and terms are examined to provide the reader with an improved understanding of the terms frequently encountered in any discussion of electronic displays.

A foundation definition is that for a lumen: the amount of flux in a solid angle of 1 steradian (sr) radiating from $1/60$ cm^2 of platinum at 2046.65°C. Luminous intensity is expressed in candles (or candelas). A candle is defined as 1 lm/sr. For extended area sources, the intensity is expressed in candles per square meter (or other unit of area).

Luminous energy is that part of radiated energy capable of evoking a visual response in the human eye. Luminous energy is generally only a small fraction of the total energy radiated by an object. Luminous efficiency is a measure of the ability of a material to convert electrical energy into luminous energy, commonly expressed as lumens per watt. Typical values are 12 lm/W for an incandescent source and 58 lm/W for a fluorescent source. At the peak response of the human eye, approximately 555 nm, under typical daylight (photopic vision) conditions, 1 W can produce 680 lm. Under dark (scotopic vision) conditions, the curve shifts toward the blue with the peak at just over 500 nm. Figure 3.7 is the relative luminosity curve that illustrates how quickly the human eye response falls off in either direction from the peak. This rapid drop in response is significant to display designers since it means that relatively large amounts of power are required to produce blue or red images that will appear as bright as green images.

Table 3.3 provides example values for commonly encountered luminous sources and luminances.

References

1. Haakenstad, L. K., "System Development and Validation Testing: 757/767 Electronic Flight Instrument System," *Proceedings of the IEEE/AIAA Fifth Digital Avionics Systems Conference*, 83CH1839-0, 19.4.5, 1983.

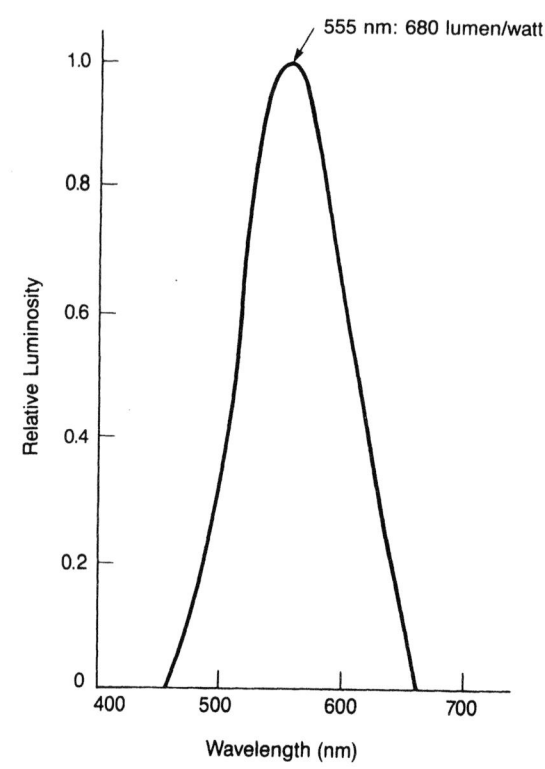

Figure 3.7 Luminosity curve.

2. Robbins, Lionel, et al., "8" × 8" Full Color Cockpit Display," *IEEE Aerospace and Electronic Systems Magazine*, v. 5, no. 9, pp. 3–6, September 1990.
3. Advisory Circular 25-11 Transport Category Airplane Electronic Display Systems, July 16, 1987.
4. Plantier, Denis, and Faivre, Francois, "An Avionics Full-Color Collimated Head Level Display," *SID 91 Digest*, Society for Information Display, p. 123, 1991.
5. MIL-L-85762A Lighting, Aircraft, Interior, Night Vision Imaging System (NVIS) Compatible, 26 August 1988.
6. Adam, Gene, McDonnell Aircraft Co., personal communication.
7. DOD-HDBK-763 Human Engineering Procedures Guide, 27 February 1987.
8. MIL-STD-203G Aircrew Station Controls and Displays: Assignment, Location, and Actuation of, for Fixed Wing Aircraft, 1 March 1991.
9. FAR 25.1303 Flight and Navigation Instruments.
10. FAR 25.1321 Instruments: Installation, Arrangement and Visibility.
11. FAR 25.1322 Instruments: Installation, Warning, Caution and Advisory Lights.
12. Weiner, Earl L., "Human Factors of Advanced Technology ('Glass Cockpit') Transport Aircraft," NASA CR 177528, 1989.
13. Dickerson, M. C., and La Saxon, V. M., "Measures of Merit for Advanced Military Avionics: A User's Perspective on Software Utility," *NATO AGARD CP-439, Software Engineering and Its Application to Avionics*, pp. 1-1 to 1-14, 1988.
14. Rosenhoover, F. Allan, "AFTI/F-16 Voice Interactive Avionics Evaluation," SAE paper 871751, 1987.

Chapter 4

Avionics Systems Essentials III: Power

The avionics designer has always assumed power would be available in whatever form necessary and that it would never be interrupted. Even if it were to be interrupted, with analog electronics it is relatively easy to design for such a possibility. However, with the digital age, life is not that simple. Because of the sensitivity of digital circuits to supply voltage variations and high-frequency noise, it is necessary for the designer to understand exactly what can be counted on from the power buses and then to design the system to operate within that envelope.

This chapter will examine two standards which apply to the performance of power systems for military and civil aircraft and provide hints on power system design. (It is recognized that power systems are not the forté of avionics systems designers; however, these hints are provided so the reader can contribute to what is frequently a very perplexing design problem that directly affects the avionics performance.)

MIL-STD-704: Aircraft Electric Power Characteristics establishes the requirements for electric power as delivered to the input of the using equipment on military aircraft. It spells out in detail what the user can expect to see in terms of normal conditions, over- and undervoltage, frequency drift, and so on.

While MIL-STD-704 describes the electrical power quality for military aircraft and how equipment is required to operate when receiving such power, RTCA document DO-160: Environmental Conditions and Test Procedures for Airborne Equipment presents comparable information for civil aircraft. DO-160 does not levy power quality requirements, per se, but instead describes a series of input power conditions and how the equipment under test must perform when receiving such power. Section 16 of DO-160 presents a definition of a given aspect of power quality and then presents a test requirement in similar but less demanding terms. Also included is a description of the test apparatus and protocols.

MIL-STD-704 Applies to Military Aircraft

As stated earlier, MIL-STD-704: Aircraft Electric Power Characteristics establishes the requirements for electric power as delivered to the input of the using equipment on military aircraft. This section will expand on and clarify these requirements and state further requirements for operation under other than normal conditions.

The steady state ac power characteristics per phase are shown below for a standard military 115-V, 400-hertz (Hz) system:

Voltage	108.0 to 118.0 V rms
Frequency	393 to 407 Hz
Maximum dc component	±0.10 V

For three phase systems:

| Phase unbalance | 3 V |
| Phase difference | 116 to 124° |

Figure 4.1 summarizes the allowable ac voltages under normal, over-, and undervoltage conditions. It is important to note that even for normal operating conditions the supply voltage is allowed to range from 80 to 180 V rms during transients that last up to 0.01 s. If the ac system nom-

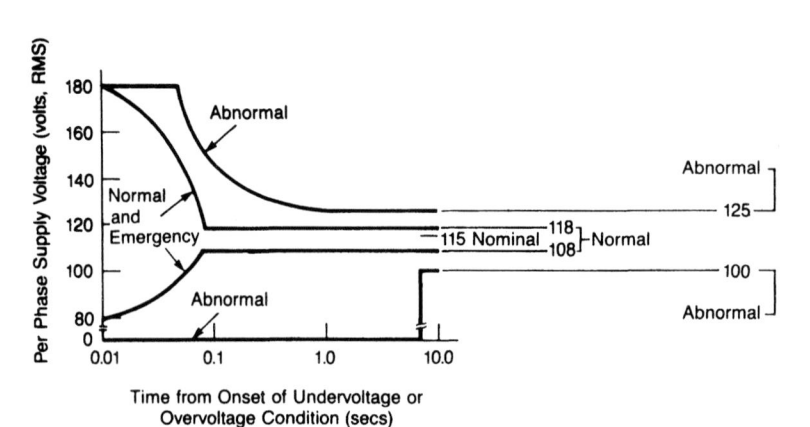

Figure 4.1 MIL-STD-704E 115-V 400-Hz power system voltages for various conditions.[1]

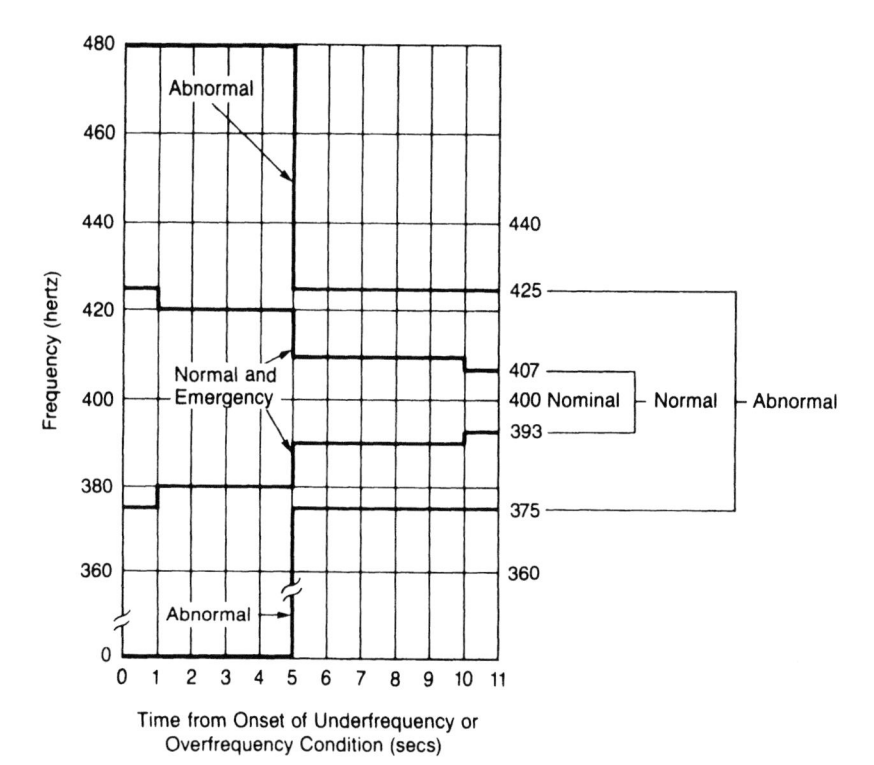

Figure 4.2 MIL-STD-704E 115-V 400-Hz power system frequencies for various conditions.[1]

inal voltage is 230 V, all of the values in Fig. 4.1 are doubled. The companion Fig. 4.2 presents the allowable ac frequency for the same set of operating conditions. Again note that the frequency is allowed to vary from 375 to 425 Hz for up to 1 s under normal conditions.

If the electrical power is nominally 28 V dc, the supply voltage is allowed to have steady state values in the range from 22.0 to 29.0 V. Ripple must be less than 1.5 V. Figure 4.3 shows the allowable ranges of voltages for a 28-V dc system under the same set of operating conditions as prescribed for ac systems. Figure 4.4 shows the same types of information for the 270-V dc case. Ripple must be less than 6.0 V.

There are two important points to be made. During undervoltage conditions, power may be completely interrupted for up to 7 s and under transfer conditions power may be interrupted for up to 50 ms followed by return to the normal curve.

MIL-STD-704 also prescribes how the user equipment must perform for all power conditions. "When supplied electric power having characteristics specified herein, aircraft utilization equipment shall provide the level of performance required by its detail [sic] specification

Chapter Four

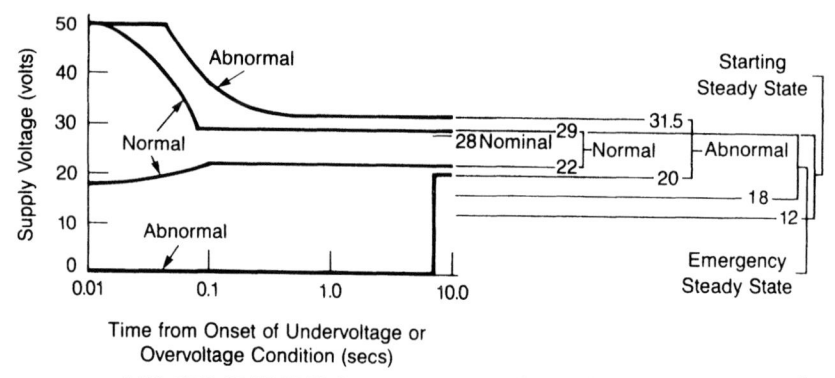

Figure 4.3 MIL-STD-704E 28-V dc power system voltages for various conditions.[1]

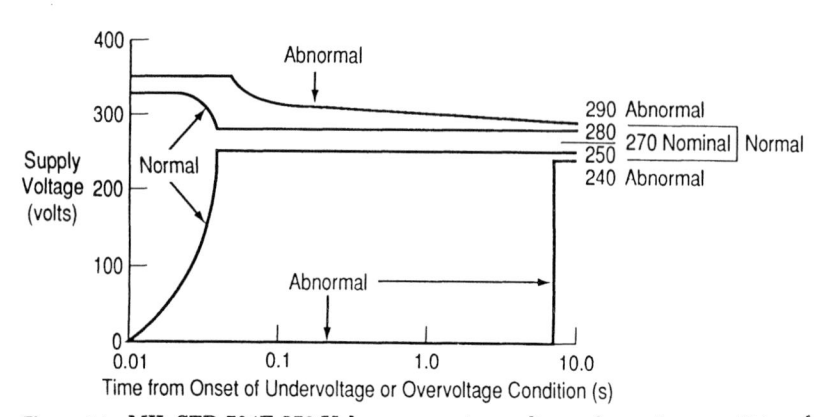

Figure 4.4 MIL-STD-704E 270-V dc power system voltages for various conditions.[1]

for each operating condition." For abnormal and transfer conditions, the using equipment may be permitted a degradation or loss of function (unless otherwise stated in its own specification), but it must not produce a dangerous or unsafe condition and must automatically recover full specified performance when power is restored to normal operating limits. When the using equipment specification requires operation in emergency or starting conditions, the equipment shall be capable of full operation when the voltage (or frequency) is within the limits shown in Figs. 4.1 through 4.4.

For voltage transients lasting less than 50 µs, equipment performance is governed by MIL-E-6051: Electromagnetic Compatibility Requirements, Systems. A discussion of this specification is given in Chap. 9 in the section dealing with electromagnetic interference.

If the equipment is supplied with more than one type of power, the loss of any one type shall not result in an unsafe condition or damage the equipment. This requirement is modified by MIL-F-9490 for flight control systems so that if any one type of power is lost, all other types shall be automatically disconnected.[1]

DO-160 Prescribes Civil Aircraft Electrical Power Quality

The previous section described the electrical power quality for military aircraft and how equipment must operate when receiving such power. This section will present the comparable information for civil aircraft. As mentioned earlier, power quality for civil aircraft avionics is prescribed in RTCA document DO-160: Environmental Conditions and Test Procedures for Airborne Equipment. DO-160 does not levy power quality requirements, per se, but rather describes a series of input conditions and how the equipment under test must perform in the presence of such power. Part A of App. B. gives the purpose of each power quality described in DO-160.[2]

Since aircraft electrical power can be either ac or dc and can be generated in several ways, section 16 of DO-160 establishes four classes of equipment. Type A equipment is intended for use on aircraft that use primarily ac power, and any dc power is derived from transformer/rectifier units. Floating a battery on the dc bus is optional. Type B equipment is intended for use on aircraft where the dc power is supplied by engine-driven alternator/rectifiers or generators, and there is a battery of substantial capacity floating on the bus at all times. (The battery serves as both a backup and a filter on the bus.) Category E is equipment powered only by ac. Type Z equipment may be used on all other types of systems covered by the document. Category Z is acceptable for use in lieu of Category A.

The steady state ac power characteristics per phase are shown below for a standard civil 115-V 400-Hz system:

Voltage	104.0 to 122.0 V (rms)
Frequency	380 to 420 Hz

For three phase systems:

Phase unbalance	3 V
	4 V (emergency power system only)
Phase difference	118 to 122°

During normal operation, power interrupts up to 1 s can be expected, depending on the type of power system. The equipment under test must resume normal operation after the interrupt. However, a re-

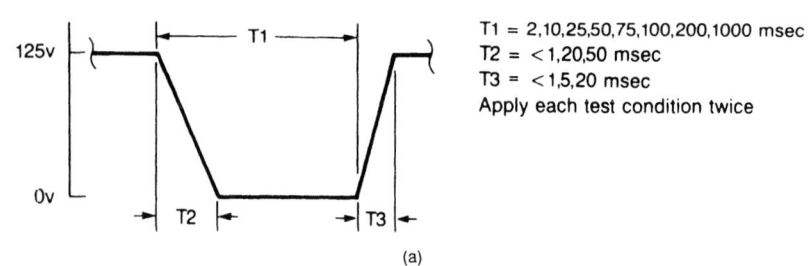

```
                              T1 = 2,10,25,50,75,100,200,1000 msec
125v   ⎤⎯⎯⎤   ⎯⎯ T1 ⎯⎯   ⎡⎯⎯⎡   T2 = <1,20,50 msec
       |   \            /    |    T3 = <1,5,20 msec
       |    \          /     |    Apply each test condition twice
       |     _____/      |
 0v    ⎦       ⎯⎯    ⎯⎯
            →| T2 |←    →| T3 |←
                      (a)
```
Figure 4.5a DO-160C ac momentary power interruption test.[2]

set may be allowed. Figure 4.5a displays the waveform that must be applied to the equipment to determine compliance with this requirement. Test condition 1 calls for T1 to be 2 ms and T2 and T3 to be less than 1 ms. For the remaining test conditions for ac powered equipment, T1 ranges from 10 to 1000 ms. T2 is either 20 or 50 ms, and T3 is either 5 or 20 ms. Equipment in Categories A and E have a total of 13 test conditions and a maximum T1 of 200 ms. Category Z equipment has a total of 15 test conditions and a maximum T1 of 1000 ms. Each test condition must be applied twice. Depending on the equipment specification, performance can be checked either during or following application of the test waveform.

During over- and undervoltage surges, as shown in Fig. 4.5b and c, the equipment is usually allowed degraded performance. However, it must automatically resume normal operation following the surge. For this test, the normal surge voltage test waveform must be applied three times.

Abnormal voltages are a substantial departure from the normal situation, and, as a general rule, the equipment is not expected to operate at full capability with such voltages. However, normal operation must be automatically resumed following the abnormal condition. Figure 4.5d shows the waveform to be applied three times in the abnormal ac voltage surge test.

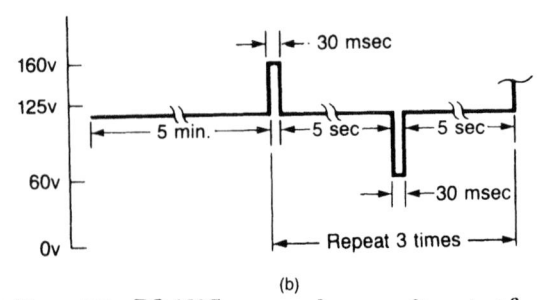

```
                    →| |← 30 msec
160v   ⎤                ⎡⎤
125v   |⎯⎯⎯⎯⎯⎯⎯⎯⎯⎯⎯⎯⎯⎯⎯⎯⎯⎯⎯⎯⎯⎯⎯⎯⎯⎯⎯⎯⎯⎯⎯⎯
       |     ⎯⎯ 5 min. ⎯⎯  ⎯5 sec→  ⎯5 sec→
 60v   |                           ⎣⎦
                              →| |←30 msec
  0v   ⎦
            ⎯⎯⎯⎯⎯ Repeat 3 times ⎯⎯⎯⎯⎯
                        (b)
```
Figure 4.5b DO-160C ac normal surge voltage test.[2]

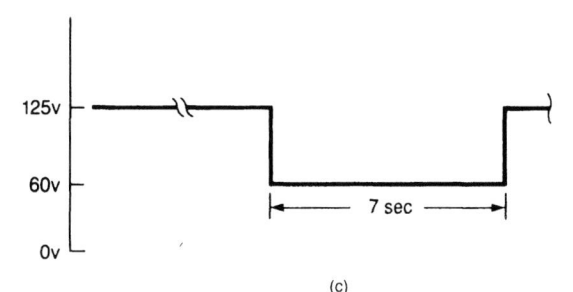

Figure 4.5c DO-160C ac momentary undervoltage test.[2]

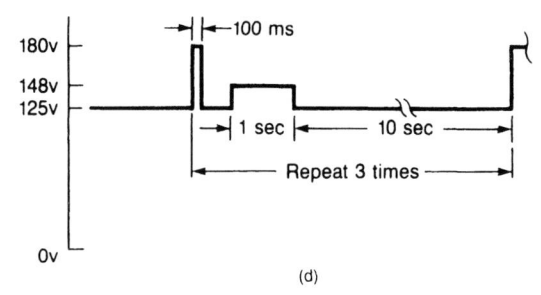

Figure 4.5d DO-160C ac abnormal surge voltage test.[2]

For 28-V dc powered equipment, many of the tests for performance under off-nominal conditions use the same general waveforms as ac powered equipment, as shown in Figs. 4-6a through d. In fact, the momentary power interruption test is the same for both types of equipment. The major differences are that the input voltage is 28 V dc, and the maximum interruption, T1, is 50 ms. For the surge voltage tests, the waveform is the same, but, of course, the voltage values are different.

Section 17 of DO-160 describes additional tests and associated test equipment and protocols to assess the performance of equipment in the presence of voltage spikes. For the voltage spike tests, the equipment under test is divided into two classes (not to be confused with the

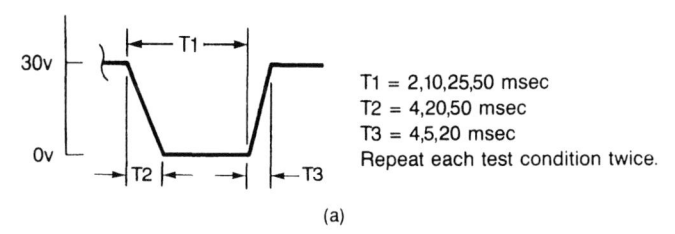

T1 = 2,10,25,50 msec
T2 = 4,20,50 msec
T3 = 4,5,20 msec
Repeat each test condition twice.

(a)

Figure 4.6a DO-160C dc momentary power interruption test.[2]

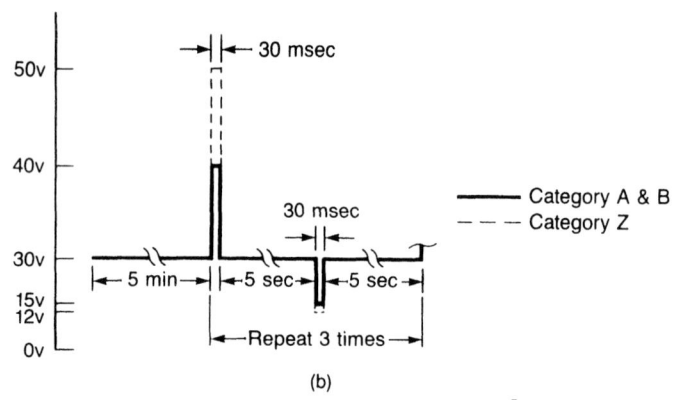

Figure 4.6b DO-160C dc normal surge voltage test.[2]

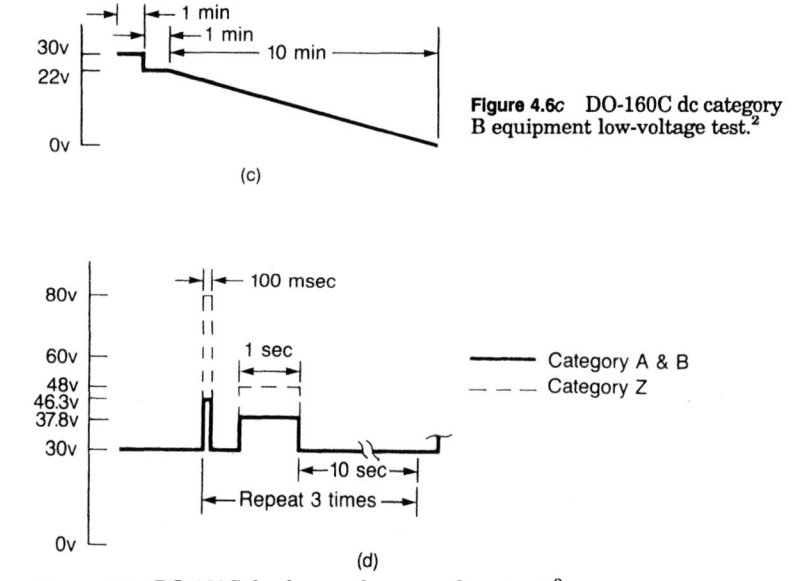

Figure 4.6c DO-160C dc category B equipment low-voltage test.[2]

Figure 4.6d DO-160C dc abnormal surge voltage test.[2]

classes used for the power input tests above): Category A for equipment which must have a high degree of protection against damage from voltage spikes and Category B for that equipment which can have a lower immunity. For Category A equipment the applied voltage spike reaches a maximum of ±600 V (open circuit conditions) in 2 µs or less and decays to 0 V in 10 µs. The spike must be applied alternately in opposite polarity at a total rate of 100 times per minute (50 spikes of each polarity).

For Category B equipment, the tests are divided into intermittent and repetitive transients. Intermittent transients of ±50 V (under load) with a rise time of 25 μs or less are superposed on the dc power input at a rate of two per second. Repetitive transients are applied at a rate of two per second for at least 2 min to both the ac and dc power inputs. The dc pulses are the same as for the intermittent case except the voltage is reduced to 20 V. The ac pulses reach a maximum of twice the rms line voltage in 2 μs and decay to 0 V by 10 μs.

Comparing the Military and Civil Requirements

Table 4.1 compares aircraft electrical power as described in MIL-STD-704 and DO-160. Figure 4.7a through c graphically compare the tabulated data. Figure 4.7a shows a comparison of the ac voltage limits and Fig. 4.7b compares the frequency limits for various operating conditions as presented in the table. Figure 4.7c compares the dc voltage requirements.

Note that for the 115-V ac case, abnormal voltage as low as 97 V per phase may occur for civil aircraft. Also note that in the case of emergency conditions for civil aircraft, a minimum voltage is not specified. However, in either civil or military aircraft for that condition the frequency may range from 360 to 440 Hz.

In the 28-V dc case, DO-160 establishes categories of power systems as discussed in the earlier section of the document. Of particular note in the table is that under starting conditions for civil aircraft, the dc voltage may drop to 10 V for up to 15 s. Since selected avionics may have to operate during engine starting, this is a particularly severe condition. DO-160 does not provide an upper limit on voltage during emergency and starting conditions.

In all cases, under normal voltage conditions, equipment is expected to operate with full capability. Under abnormal voltage or frequency conditions, the equipment operational capabilities are spelled out in the equipment specifications.

Tips on Power System Design

Although the power system design is not the responsibility of the avionics designer, it is very important since it provides the life blood for the avionics. Consequently, this section is included to present useful power system design hints that may make the avionics designer's task a little easier.

TABLE 4.1 Comparison of Aircraft Electrical Power: MIL-STD-704E and RTCA DO-160C

Voltage	MIL-STD-704	DO-160*
\multicolumn{3}{c}{AC voltage}		
Normal	108.0–118.0 Transients 180 V for 10 ms 80 V for 10 ms	104.0–122.0 Cats. A, E, Z Interrupts ≤200 ms A, E ≤1 s Z Surges 160 V for 30 ms 60 V for 30 ms
Over- and undervoltage	100.0–125.0 Transients 180 V for 50 ms 0 V for 7 s	97.0–134.0 Undervoltage 60 V for 7 s Surges 180 V for 100 ms 148 V for 1 s
Emergency	108.0–118.0 (steady state)	NA
\multicolumn{3}{c}{AC frequency (hertz)}		
Normal	393–407 Transients 375–425 for 1 s 380–420 for 5 s 390–410 for 10 s	380–420 Cats. A, E, Z
Abnormal	374–425 Transients 0–480 for 5 s	(Same as normal)
Emergency	360–440 (steady state)	360–440 Cats. A, E, Z
\multicolumn{3}{c}{DC voltage}		
Normal	22.0–29.0 (28.0 nominal) Transients 50 for 12.5 ms 18 for 15.0 ms	22.0–29.5 (27.5 nom.) A, Z 24.8–30.3 (27.5 nom.) B Interrupts ≤200 ms A ≤50 ms B ≤1 s Z Surges 40 for 30 ms A, B 15 for 30 ms A, B 50 for 50 ms Z 12 for 30 ms Z

Avionics Systems Essentials III: Power 87

TABLE 4.1 Comparison of Aircraft Electrical Power: MIL-STD-704D and RTCA DO-160 (*Continued*)

Voltage	MIL-STD-704	DO-160*
	DC voltage	
Over- and undervoltage	20.0–31.5	20.5–32.2 Cats. A, Z
		22.0–32.2 Cat. B
	Transients	Undervoltage
	50 for 50 ms	12 for 7 s
	0 for 7 s	Surges
		46.3 for 100 ms A, B
		37.8 for 1 s A, B
		80 for 100 ms Z
		48 for 1 s Z
		Low voltage
		22–0 in 10 m B
Emergency	18.0–29.0 (steady state)	18.0 Cats. A, Z. 20.0 Cat. B
Starting	12.0–29.0 (steady state)	10 for 15 s

*Category A: Mainly ac power with dc from transformer/rectifier units; B: Solely dc power with a large battery on the bus; E: Solely ac power; Z: Mainly ac power with dc from variable speed generators.

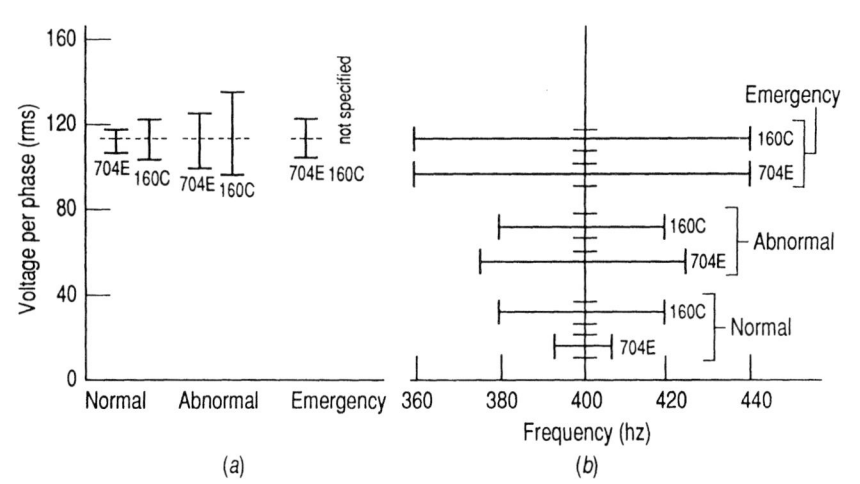

Figure 4.7 Comparison of ac (*a*) voltage and (*b*) frequency: MIL-STD-704E vis-à-vis DO-160C.

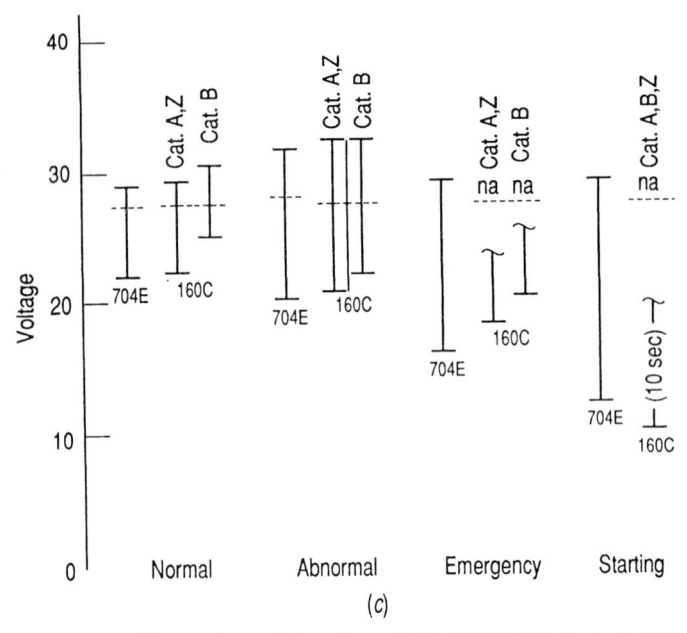

Figure 4.7c Comparison of dc voltage: MIL-STD-704E vis-à-vis DO-160C.

The keys to sound design in power systems are the same as for avionics: classification and segregation of function, redundancy, and dissimilar design approaches.[3]

The electrical loads on aircraft must be classified as critical, essential, or utility, and buses with appropriate levels of power quality and redundancy must be established for each type of load. Critical loads are typified by the flight control system and the cockpit displays. Essential loads for all types of aircraft include anti- and de-icing and the Environmental Control System (ECS). For military aircraft the mission avionics, such as electronic or antisubmarine warfare, should be treated as essential loads. Typical utility loads on commercial transports are the galley, landing gear, and passenger entertainment system.

Critical electrical loads, by definition, are the most important loads on the aircraft and require the most attention. Fortunately, critical loads are usually a small fraction of the total load. Quadruple redundant supplies from dissimilar sources are strongly recommended for critical loads. As a general rule, these loads are supplied by two redundant main buses, a backup bus, an emergency bus, and/or batteries. The main buses are fed from different generators on different engines if it is a multiengine aircraft. The backup can be from an auxiliary power unit or from a hydraulically driven generator. For emergency power, ram air turbines do a good job, and wing-tip-mounted, vortex-driven turbines are also another source.

Batteries are always a possible last-ditch power source. However, there are significant maintenance ramifications in their use. They offer the capability of providing instant power to support the load while other emergency sources are brought on line. The selection of equipment to be powered by batteries must be strictly limited since the larger the emergency load, the larger the batteries need to be to support it. Batteries must be constantly recharged and have a provision for conveniently checking their charge state.

Essential loads usually receive at least triple redundant power, being supplied by two main buses and at least one backup bus. Since essential loads can generally be selectively operated, instantaneous power on the essential bus can be markedly reduced.

Finally, utility buses are usually dual redundant and supplied by two main generators. Loads on the utility bus can be operated in a completely selective way so that if less than normal power is available, most power-demanding activities can proceed, albeit at a slower pace.

There are many electrical system design tips that have been formulated over the years, which are summarized below:

The routing of power buses, like data buses, should be dispersed throughout the aircraft to minimize possible damage from a single hazardous event.

Switching from one ac source to another without severe transients can be achieved through microprocessor-controlled synchronization prior to putting the second machine on line.

The $N1$ stage of a jet engine can be used for emergency electrical power generation by installing small magnets in the tips of the turbine blades and installing small coils in the engine housing. This approach has the advantage that power can be generated even when the engine is only windmilling, and there is no delay in connecting the source. Small turbine generators for emergency power can be powered by gases produced from burning hydrazine, a monopropellant.

Batteries are an obvious choice to provide emergency power. However, they should be centrally located so they can be easily checked and maintained. There should be provisions for trickle charge at all times, both in flight and on the ground.

Small hydroelectric generators that convert hydraulic power to electrical power can be remotely located at sensors and microprocessors.

One useful technique in bus design is the dc link where the ac generator output is rectified for transmission on the bus and then inverted at the using equipment. This approach allows true gapless power since batteries can be on the bus as a hot backup at all times. A disadvantage is the switching of large currents. However, solid state power controllers are being developed that can handle the large loads.

Variable voltage/variable frequency is an approach to power systems that offers simplicity and low weight. In this case, the ac generators are directly coupled to the engine and the output power frequency and voltage are proportional to the engine speed. Only the power required for critical loads is conditioned. Most of the power is used in the unconditioned form by the large loads such as galleys, ECS, and landing gear.

References

1. MIL-STD-704E: Aircraft Electric Power Characteristics, 1 May 1991.
2. DO-160C: Environmental Conditions and Test Procedures for Airborne Equipment, RTCA Inc., 1989.
3. Ideker, J. B., "Electrical Power Integration with Redundant Future Fighter Avionics, Lear-Siegler, Inc. Astronics Division, 1983.

Chapter

5

Fault Tolerance

Fault tolerance is the ability of a system to continue satisfactory operation in the presence of one or more nonsimultaneously occurring hardware or software faults. Fault tolerance becomes especially significant when the system performs a flight critical or flight essential function, as defined by Federal Aviation Regulation (FAR) Part 25.1309: Equipment, Systems and Installation or by MIL-F-9490: Flight Control Systems—Design, Installation and Test of Piloted Aircraft, General Specification for. In brief, FAR 25.1309 specifies a probability of failure for a flight critical system of $<10^{-9}$ per flight hour. MIL-F-9490 levies a requirement for a probability of failure resulting in the loss of the aircraft of $<10^{-7}$ per flight hour.

To meet these and other demanding performance requirements, a fault-tolerant system generally uses both types of fault tolerance (hardware and software) and has the capability for automatic, dynamic reconfiguration of the system. Different levels of redundancy (dual, triple, or quadruple) are used, depending on the level of criticality and, therefore, on the allowable probability of failure. Redundancy extends to all hardware elements, such as processors, sensors, actuators, and data buses, and to the software.

When dealing with redundant hardware and software, there is the question of whether the redundant elements should be similar or dissimilar. There are powerful arguments for either choice. Similar redundancy simplifies the design process and reduces costs and programming and verification activities, but it does not protect against generic errors. Conversely, dissimilar redundancy makes the design and the programming and verification activities more complex, lengthy, and expensive but provides what is generally agreed to be a substantial increase in protection against generic faults.[1] Despite the cost and complexity, the dissimilar approach is usually chosen to further ensure meeting the system reliability goals.

Reconfiguration is the dynamic reallocation of redundant elements by executive-level software in response to failures or changes in the aircraft mission or condition. Reconfiguration and anonymity of the active elements to the user are considered basic attributes of advanced fault-tolerant systems. In case of a failure, the failed unit is automatically switched off line and its functions are assumed by a spare, fault-free unit, or different fault detection schemes are invoked.

Central to all of the fault-tolerance principles described above is the issue of fault detection. It is obvious that before any fault-tolerance scheme can be invoked, a fault must be detected. There are several approaches to fault detection: replication (triple or higher) and voting, duplication and comparison, and self-checking. In replication and voting, a highly fault-tolerant voting circuit compares the values from multiple processors computing the same parameter, and if one of the values does not agree with the others, the value is ignored and the processor that generated the suspect value is switched off line. Based on the degree of fault tolerance required in the system, a replacement processor can be brought on line or the system can revert to a lower level of replication or to the duplication and comparison mode of operation. The failed processor may, if so designed, then execute a self-diagnostic check and, if no permanent faults are found, return to active status.

Fault-Tolerant Hardware and Systems

The most common application of fault tolerance is to flight critical flight control systems as noted in the introduction to this chapter. Figure 5.1 shows an Allied Signal Bendix/King generic electronic flight

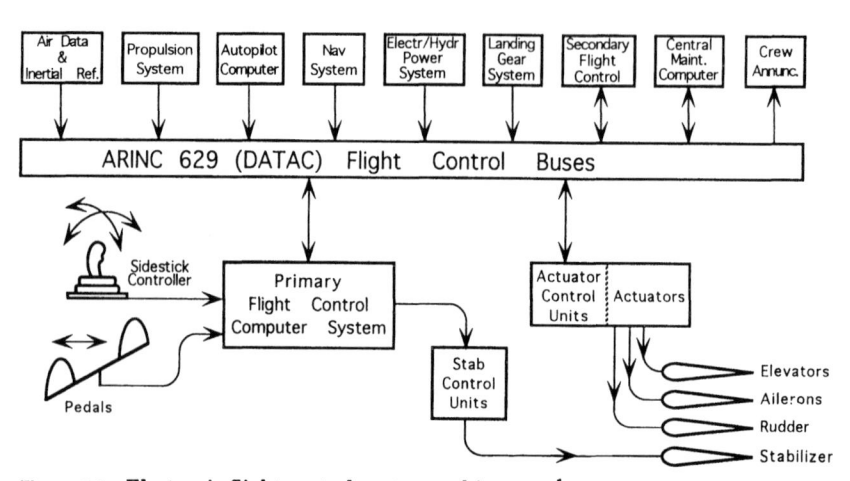

Figure 5.1 Electronic flight control system architecture.[1]

control system architecture for a commercial transport. Note the use of ARINC 629 buses as the means of integrating the system and the central, essential role of the primary flight control computer system (PFCS). The PFCS uses information from other aircraft systems transmitted over the ARINC 629 buses along with inputs from the sidestick controller and pedals to generate control signals for the flight control surface actuators.[1]

Figure 5.2 is a detailed view of the PFCS. There are left and right primary flight computers (PFCs), each with triple redundant lanes for a total of six lanes. Within each PFC the lanes have dissimilar hardware and software; however, from PFC to PFC the corresponding lanes are identical; that is, the left lane of the left PFC is identical to the left lane of the right PFC.

Figure 5.3 shows the details of a single lane of the MAFT. Each lane has two major sections, the operations controller (OC) and the flight control electronics (FCE). The OC and the FCE are well partitioned from each other to improve the certificability of the PFCS. The OC is a special very large-scale integration (VLSI) processor that executes the executive overhead tasks including interlane communication, synchronization, data voting, error detection, task scheduling, and reconfiguration. Allocating the executive functions to the OC frees up the FCE to handle only flight control tasks. Additionally, this partitioning of functions enables hardware and software specialists to focus their attention solely on either the OC or FCE.[2]

The FCE contains an applications processor (AP) and an input-output processor (IOP). The AP handles the control laws and secondary tasks such as watchdog timers and operational monitoring. The IOP interfaces with selected analog and digital devices and with the

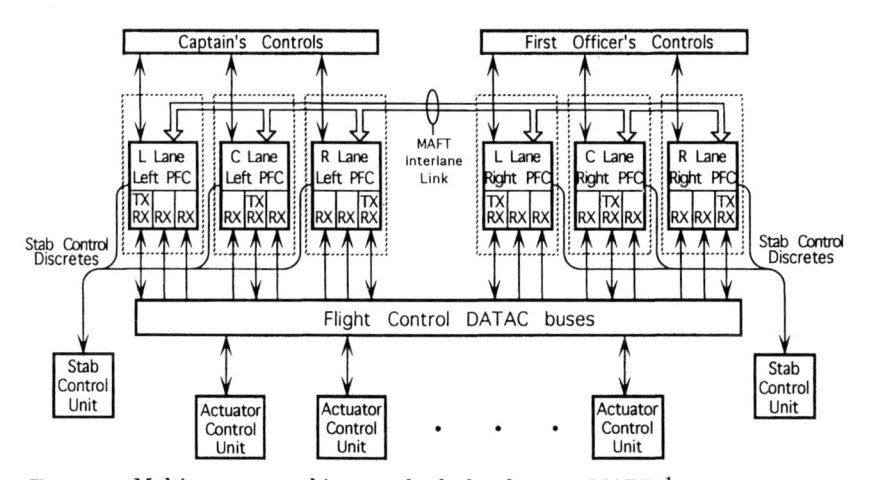

Figure 5.2 Multicomputer architecture for fault tolerance (MAFT).[1]

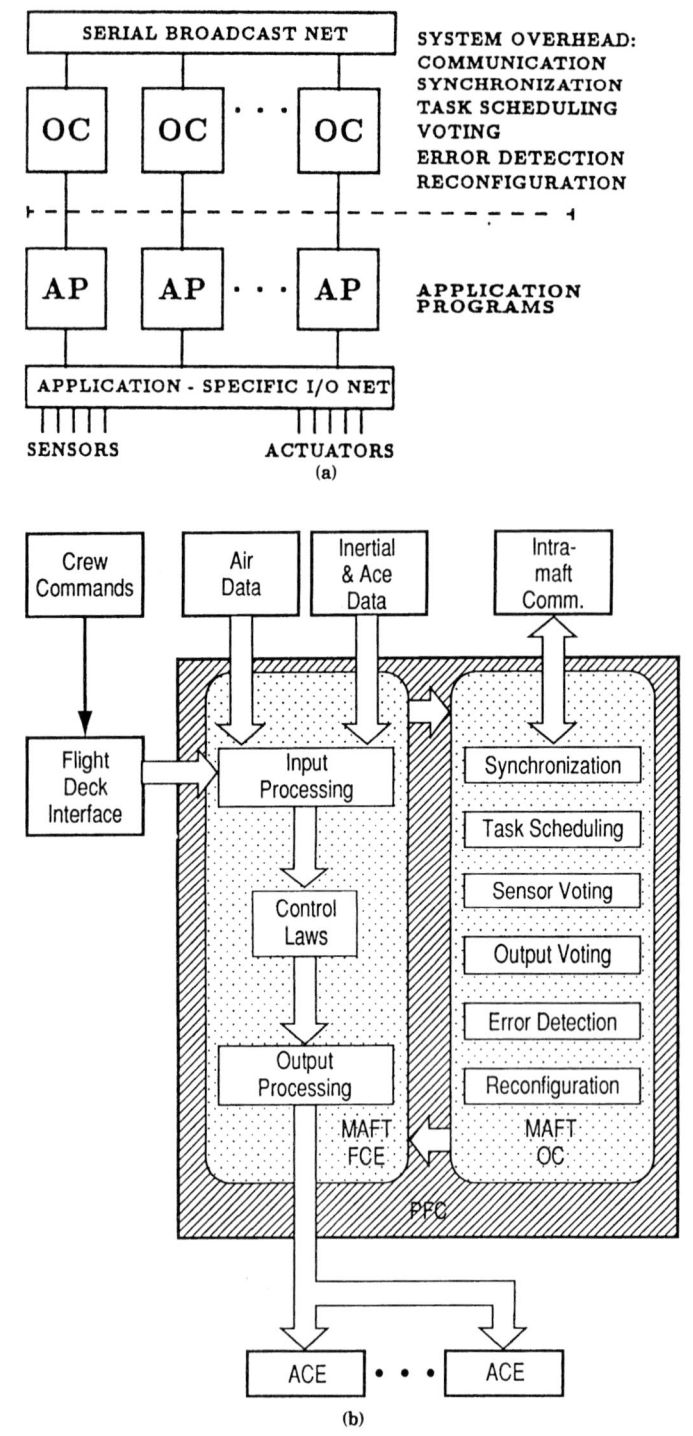

Figure 5.3 MAFT lane architecture (ACE = actuator control electronics).[1]

ARINC 629 flight control buses. An IOP receives data from all three flight control buses but transmits data on only one of them.

Another approach to triplication and voting, developed by GEC Avionics, is shown in Fig. 5.4. In this concept there are three PFCs, each receiving data from all three flight control buses but only transmitting data on one bus.[3]

Details of a PFC are shown in Fig. 5.5. The subfunction elements of a PFC are, in sequence, peripheral links, processor links, and ARINC 629 links. Each link is quadruple redundant with, in a given PFC, identical hardware and software. Within a PFC any peripheral can be connected to any processor link, and any processor link can be connected to any ARINC 629 link. A PFC can remain operational as long as a single path can be completed through the links; thus, only one of the four redundant units in any link has to be functioning for a PFC to be operational. Quadruple redundancy in each link leads to extremely high reliability. The probability of flight dispatch, which requires two healthy paths in two PFCs and one healthy path in the remaining PFC at the beginning of the flight, is predicted to be 0.95 at 38,100 operating hours. The mean time between maintenance actions is estimated to be 92,700 operating hours.[3]

Dissimilar hardware and software are used in the three PFCs. The left PFC uses an Inmos Transputer T414 processor programmed in

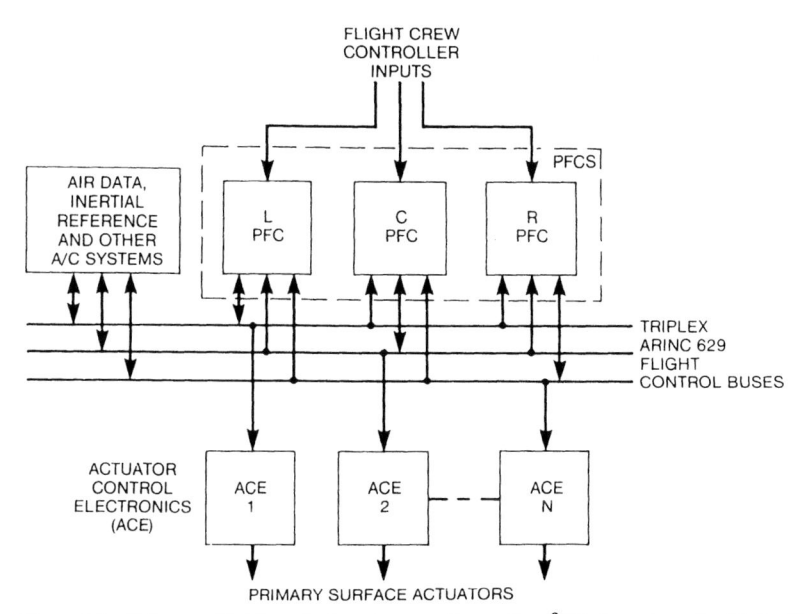

Figure 5.4 Primary flight control system configuration.[3] (© *American Institute of Aeronautics and Astronautics; reprinted with permission*)

96 Chapter Five

```
Peripheral          Processor           Datac
  Links              Links¹             Links
```

| Analogue and Discrete Interfaces | Processor SCI ―I/O― SCI Memory | DATAC Interfaces SCI Processor Selector |

Note:
1) Inter link interconnection shown for one link type only.
2) SCI-serial communication interface.

Figure 5.5 Primary flight control system configuration.[3] (© American Institute of Aeronautics and Astronautics; reprinted with permission)

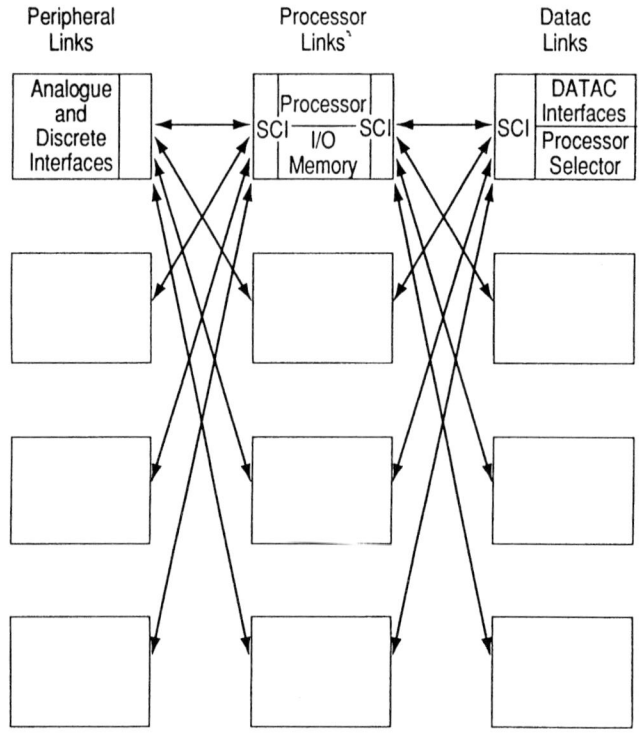

Occam, the center PFC uses a Motorola 68020 processor programmed in Ada, and the right PFC uses an Intel 80386 processor programmed in C. Each PFC fits into a passively cooled 10-modular-concept-unit (MCU) LRU.

Figure 5.6 shows an approach to triplication and voting developed by the Honeywell Co. Air Transport Systems Division.* This dual-channel flight control computer (FCC) has six microprocessors, three in each channel. Within a channel central processing units (CPUs) 2 and 3 are identical processors with identical software. CPU 1 is a different processor with different software developed by a separate team

---

*Until 1988 the Honeywell Co. Air Transport Avionics Division was the Sperry (Co.) Flight Systems—Commercial Division in Phoenix, Arizona.

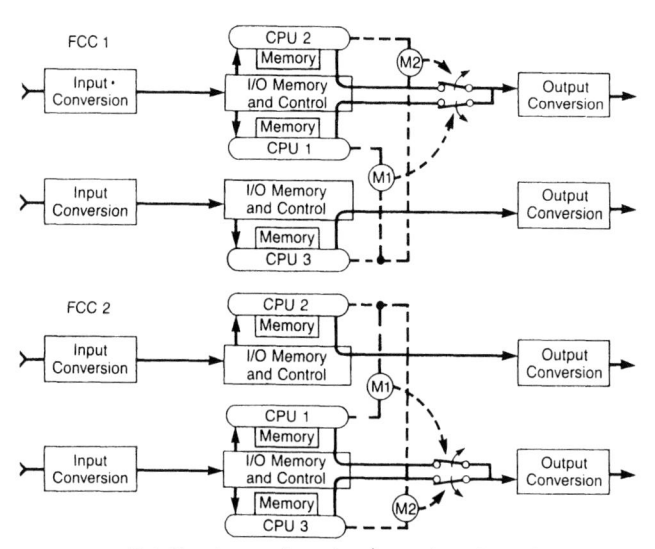

Figure 5.6 Triplication and voting.[4] (© *American Institute of Aeronautics and Astronautics;* reprinted with permission)

of programmers than the team that developed the software for CPUs 2 and 3. In the upper FCC channel, CPU 1 shares input, output, memory, and control with CPU 2, while in the lower channel CPU 1 shares similar functions with CPU 3. M1 and M2 are comparison monitors that compare the output of CPU 1 with the unpaired processor and the outputs of CPU 2 and CPU 3 with each other, respectively.[4]

CPU 1 is the primary processor and performs both essential and critical functions (as defined by FAR 25.1309). CPU 2 and CPU 3 are secondary processors that perform only critical functions. The three types of processors used in this FCC are an Intel 80386 programmed in PLM, a Motorola 68020 programmed in C, and a Sperry SDP 185 programmed in Pascal.

In case of a CPU 1 failure (detected by monitor M1), the top lane in FCC 1 remains connected to CPU 2, and the bottom lane in FCC 2 remains connected to CPU 3. Thus, both FCCs remain engaged. In the case of a CPU 2 failure (detected by monitor M2), CPU 1 remains connected in the top lane of FCC 1, but CPU 2 and CPU 3 are both disconnected in the bottom lane of FCC 2. Thus, FCC 2 is shut down. A similar situation occurs for a failure in CPU 3 except that the CPUs disconnect from the top lane of FCC 1, and FCC 1 is therefore shut down.

The fault-tolerant architecture in Fig. 5.6 shows only dual FCCs. Obviously additional FCCs could be added using the same CPU and monitor building blocks to achieve any level of reliability required for a flight control system.

In the duplication and comparison mode, two processors compare their outputs with each other, and if they do not agree, the pair of processors collectively drop off line and begin self-diagnostic routines. If each processor passes its self-check, it can return to the active state and pair with another processor, either its previous mate or another, and resume processing.

The concept of self-checking processing pairs has been developed and proven by the Honeywell Co. Systems and Research Center. Fig. 5.7 shows two typical self-checking pairs. There are two identical halves in each pair with each half containing an applications processor (AP) and a bus interface unit (BIU). Although both BIU and AP pairs in fact contain both types of monitors, to clarify the description of the operation, the left pair shows the output monitors and the right pair shows the input monitors. The output monitors compare the output of the two APs, and if they disagree, the AP pair is disconnected from the bus. In a similar vein, the input monitors compare the input values from the two buses, and in case of a miscompare, prevent the data from reaching the APs.[5] This duplication and comparison concept is used in the ARINC 659 backplane data bus for checking both transmitted and received data.

Self-checking processors are a simple concept, yet demanding in implementation. A self-checking processor can detect an error within itself through reasonableness checks on its intermediate and/or final re-

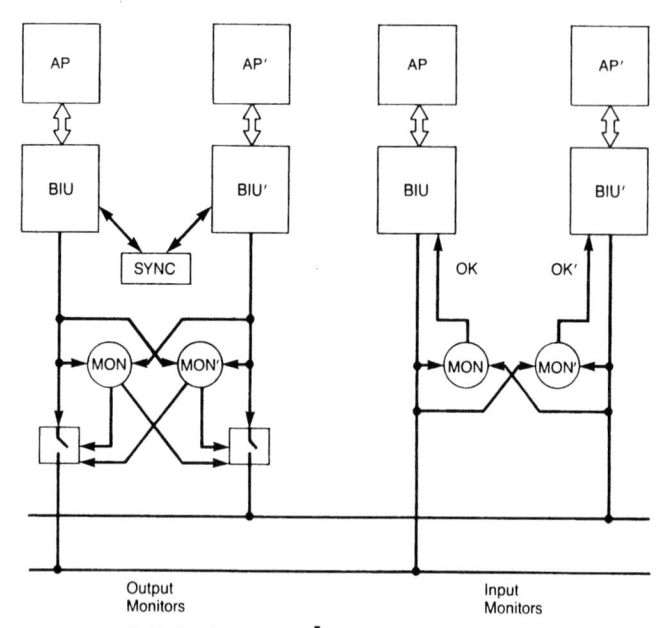

Figure 5.7 Self-checking pairs.[5]

sults without reference to other processors. In the case of an error, it will simply switch itself off and may, if so programmed, automatically bring a spare processor on line as a replacement.

Fault-Tolerant Software

Flight critical systems require fault-tolerant software to complement the fault-tolerant hardware. Many of the concepts for fault-tolerant hardware, such as similar and dissimilar redundancy and standby sparing, have parallels in fault-tolerant software. As the architect of a digital avionics system, the designer must use software design concepts as adeptly as hardware concepts. While this section focuses on possible software design concepts, Chap. 10 will describe how these concepts are coded and validated for flight systems.

Fault-tolerant software falls into three categories: multiversion programming, recovery blocks, and exception handlers. All of these techniques are subject to error if the software specification is incorrect. The importance of beginning the software design process with an accurate and complete software specification cannot be overemphasized. It is essential that the designer, the programmer, and other cognizant persons review the software specification to verify its correctness. An error in the software specification will probably produce an error in the software, one that may not be found, even in exhaustive testing, but may later cause catastrophic failure of the system.

Multiversion, or *N*-version, programming requires the development of two or more versions of a program that performs a specific function described in the software specification. These different program versions should be developed by separate software teams and may even be designed to operate on different processors. They accept a common input from an executive-level program which, in turn, also compares the results of the different versions to detect faults.

N-version program development requires careful attention to several issues. The first is the need for specific comparison points such as input and output values. Implementing comparison schemes, in turn, implies the need for synchronization among the versions so comparisons can be made without excessive delay. Finally, the designer must decide whether the versions are to be executed in parallel or sequentially. Clearly the trade-off here is between minimum hardware and slower execution (sequentially) or more hardware and maximum speed (parallel).

Recovery blocks are another concept in fault-tolerant software. Acceptability checks are made on the results from a primary version of a program. If the results fail the acceptability checks, an alternate version of the program that is different from the primary version is invoked, and the process of computation and acceptability checks is repeated. If no alternate version produces an acceptable result, the software block is judged to have failed.

Comparing the results of two or more programs, as used in N-version programs, is preferable to performing acceptance tests, as in recovery blocks, because comparisons can be performed quickly, simply, and with fewer errors than acceptance tests.

Given that a fault has occurred and has been detected, it now must be recovered from. Recovery can be either backward or forward. Backward recovery is exemplified by recovery blocks where, in case of a fault, the executive software reinitializes the program using the same input values as used in the previous cycle and attempts to execute the program again. Forward recovery is demonstrated in N-version programming where the outputs are compared and erroneous values, generated by faulty software, are ignored and only the correct value is passed to the user. Thus, even though a fault may have occurred, recovery has been made without any loss of time.

Another form of fault tolerance in software is run-time assertions. Watchdog timers check the time for a block of code to be executed, and if the code is not completed within the prescribed time, it is assumed an error has occurred. Run-time assertions are relatively straightforward to implement since the time to correctly execute a block of code can be easily established.

Hardened kernels of software are small, minimum capability blocks of code that perform critical functions. These blocks of code must be as simple as possible to achieve a specified function. They must have been exhaustively tested and exercised to ensure that few, if any, faults remain.

An emerging concept that uses software in lieu of hardware replication to achieve fault tolerance is analytical redundancy. In the case of a faulty sensor, analytical redundancy combines data from the remaining functioning sensors with data from other sources in the aircraft in algorithms that compute the most probable value from the failed sensor. This computed value is then used in the same ways as a value from a functioning sensor. An equivalent concept can be applied to flight control actuators and surfaces where, if an actuator fails or a control surface is lost, the remaining functioning actuators and surfaces can be combined in a way to offset the loss. Analytical redundancy and its companion concept for actuators are two of the cornerstones of reconfigurable flight control systems.

References

1. Darwiche, A. A., and Doerenberg, F. M. G., "Application of the Bendix/King Multicomputer Architecture for Fault Tolerance in a Digital Fly-by-Wire Flight Control System," *Proceedings of MIDCON 88*, IEEE, 1988.

2. Walter, C. J., "MAFT: An Architecture for Reliable Fly-by-Wire Flight Control," AIAA paper no. 88-3902-CP, 1988. Copyright American Institute of Aeronautics and Astronautics; reprinted with permission.
3. Hills, A. D., and Mirza, N. A., "Fault Tolerant Avionics," *AIAA/IEEE 8th Digital Avionics Systems Conference,* AIAA paper no. 88-3901-CP, 1988. Copyright American Institute of Aeronautics and Astronautics; reprinted with permission.
4. Yount, L. J., "Architectural Solutions to Safety Problems of Digital Flight Critical Systems for Commercial Transports," AIAA paper no. 84-2603-CP, 1984. Copyright American Institute of Aeronautics and Astronautics; reprinted with permission.
5. Driscoll, K., "Multi-Microprocessor Flight Control System," *Proceedings of the IEEE/ AIAA Fifth Digital Avionics Systems Conference,* 83CH1839-0, 11.3.1–7, 1983.

Chapter

6

Maintainability and Reliability

Maintenance is what is done to an avionics system after it enters service to repair those things that were never supposed to fail. Perhaps this is an overstatement, perhaps it is not. The point is that after a system enters service, one of its most important attributes is how it performs in terms of maintenance. Maintenance is a major factor in operating costs and, therefore, life cycle costs, so there is continuing pressure on the designer to deliver a system with minimum cost and maximum productivity maintenance.

This chapter discusses the attributes of digital avionics systems that receive high marks for ease of maintenance. The chapter opens by discussing design tips for easy maintenance. Systems that are easy to work on turn out that way because maintenance was a design driver from the beginning. Practically speaking, maintenance enhancements cannot be added after the equipment is built or even designed in after the design is nearly complete.

As digital avionics and integrated systems architectures become more widespread, the addition of built-in test equipment (BITE) and centralized fault display systems (CFDSs), also increases. This chapter examines these topics and shows how they can be applied for easier maintenance.

On the ground side, there are many benefits to be realized from automated test equipment (ATE). After the BITE and the CFDS have isolated and identified a fault in a line replaceable unit (LRU), the ATE takes over and further identifies the precise nature of the fault, generally including the exact component that failed. ATE can also help in generating and debugging testing software.

One of the earliest accomplishments in digital avionics by the airlines was the development of the Abbreviated Test Language for Avionics Systems (ATLAS). It proved to be very popular and was used on many systems, so it was consequently renamed Abbreviated Test Language for All Systems, which leaves the ATLAS acronym unchanged. This chapter presents the major features of this versatile and well-known language.

Reliability is one of the most important attributes of an avionics system and one that most people claim to understand, at least on the surface. It is often near the top of the list of "ilities" used to characterize military avionics.

On the other hand, reliability prediction is a very arcane skill that few people correctly understand and effectively practice. The traditional approach to reliability prediction has had, in the judgment of some persons, only limited success. In many cases the predicted reliability of an LRU is much worse (lower) than the actual reliability achieved in service. This large difference between predicted and actual reliability has long been recognized and in recent years has begun to receive some attention. The last section in this chapter takes a close look at the issues of reliability and reliability prediction and presents some thoughts intended to stimulate improvements in reliability prediction methods.

Maintenance: Who, When, Where?

The avionics system designer must ask three questions about system maintenance once it enters service: Who will be doing the work, when will it be done, and where will it be done?

The skill level of maintenance personnel is a major consideration in the design of the avionics system. If maintenance will be performed by unskilled and/or inexperienced personnel or by persons not fluent in the language of the maintenance literature, a design that is straightforward and easily understood, with built-in test equipment, procedures, and associated displays will need to be developed. The procedures and maintenance data interpretation must be highly structured and clearly stated. Although generalizations can be dangerous, it is probably safe to say that the designer can usually assume the experience levels of military maintenance personnel will be substantially less than those of their civilian counterparts. This difference in skill level is due to a much higher military technician turnover rate and the fact that the civilian technicians are often older and have had prior experience in military service.

The second question the designer must ask is, When will the maintenance be performed? Provisions for maintenance must take into consideration the amount of time required for a maintenance procedure. Reducing the time required to replace and check out an LRU will yield an equal reduction in sortie turnaround time for military aircraft. Depending on the mission of the aircraft, the avionics system architecture could be designed to allow the aircraft to be dispatched with an inoperative unit that could be repaired during overnight or other scheduled maintenance. Operation Desert Storm confirmed the possible need for avionics to be designed to operate almost continuously for up to, say, 30 days without maintenance.

The third and final maintenance question to be asked during the early system definition phase is, Where will the maintenance be performed? The distribution of maintenance tasks between the flight line (or flight deck) and the maintenance depot must be established early in the design. This distribution will have a major influence on the design of built-in test equipment and accessibility to the LRUs. Historically military avionics maintenance has been performed at three levels: flight line, wing avionics shop, and depot. More recently, in the case of the F-22, maintenance is being structured to two levels: flight line and depot. This restructuring is designed to reduce operating costs and expedite deployment.

Designing for Easy Maintenance

Designing, manufacturing, testing, and installing digital avionics is only the beginning. After the equipment enters service, it will experience many flight hours and many years of operation punctuated by maintenance to keep it working properly. Avionics designed for easy maintenance are major contributors to increasing the mission effectiveness of the aircraft while reducing operating costs.

Maintainability must be a major driver from the beginning in designing digital avionics. This claim is supported by the fact that FAA Advisory Circular 25.1309-1A System Design Analysis requires approval of maintenance procedures when discussing certification of flight critical systems. Additional support for this claim can be found in MIL-STD-470A Maintainability Program for Systems and Equipment, which calls for a series of maintainability tasks to be performed for any new major system.[1]

For military avionics, maintainability is defined as the measure of the ability of an item to be retained in or restored to specified condition when maintenance is performed by personnel having specified skill levels, using prescribed procedures and resources, at each prescribed level of maintenance and repair.[2]

This section presents some practical design tips for enhanced maintainability beginning with one that is frequently overlooked or ignored, which is make maintenance manuals, procedures, and equipment easy to use and understand. As these items take shape, keep in mind who is going to use them and where they are going to be used.

Poorly written manuals and poorly designed test procedures and equipment encourage shortcuts and impromptu tests that are hazardous to the avionics, the aircraft, and maintenance personnel.

Designing for easy maintenance includes using standard parts. One-of-a-kind and other hard to find parts will make maintenance difficult, if not impossible, especially after the equipment has been in service for several years. Furthermore, infrequent exposure to unique parts and circuit designs makes it difficult for maintenance personnel to remember procedures and acceptable equipment performance limits.

Accessibility is a major factor in ease of maintenance, both on the flight line and in the shop. Flight line accessibility, especially in the crowded conditions of an aircraft carrier flight or hangar deck, requires that the LRU by easy to get to and to remove. There should be ample clearance around the LRU. Removal should not require the removal of other LRUs, the use of special tools, or maintenance personnel getting into a potentially dangerous position. Access to high mean-time-between-removal (MTBR) units can be less convenient but still should not require a long time or special tools.

Each hardware component should be clearly labeled and each function should be implemented on a single replaceable unit. In the case of redundant elements, each should be individually testable. LRMs should be keyed to prevent inadvertent insertion into the wrong slot of the cabinet.

Accessibility is also an important factor in shop maintenance and repair. Test points should be readily accessible without the use of extender boards or special tools. Access to all test points should be possible with the LRU in its normal operating position.

Although such a procedure is never recommended, as a safety measure the LRU should be designed to allow its removal while powered up without being damaged. If the LRU uses electrostatic-discharge (ESD) sensitive devices, protection should be built in to the LRU to prevent possible damage when removing or installing it. The LRU should have a label on the front panel alerting personnel to the ESDs inside.

If the avionics are designed to DOD-STD-1788 Avionics Interface Design Standard or ARINC 600 Air Transport Avionics Equipment Interface, there should be no connectors on the front of the LRU. Front panel connectors mean added weight and complexity and a possible reliability penalty. Any troubleshooting that requires access to internal buses or test points should be done in the shop using the standard connectors on the rear panel.

A constant source of *could not duplicate* (CND), *retest O.K.* (RTOK), and *unverified removals* (UR) is differences between the LRU BITE and the ground-based test equipment. Test sequences and limits in both sets of equipment must be identical. Another possible source of CNDs, RTOKs, and URs are the environmental effects. BITE, discussed later in this chapter, should include measurements of the environment to the extent possible whenever a fault occurs.

The high price of RTOKs and other maintenance nuisances is shown in Fig. 6.1. The trend shown in the figure is based on the fact that testing a unit that is good is at least as expensive as testing a bad unit since several tests may be required to confirm the satisfactory performance of the good unit, and any testing, whether on a good or bad unit, has logistics and administrative overhead.

BITE and CFDS Are Big Helps

BITE is a powerful maintenance tool that takes advantage of the intrinsic capabilities of digital avionics. It is the hallmark of a system designed for easy maintenance, and it pays big dividends. As BITE has become used in more LRUs and as integration has become the central theme in avionics architectures, there has been a trend toward the development of CFDSs that display and store fault data for all BITE-equipped LRUs on the aircraft. BITE and CFDS have rapidly become indispensable maintenance aids.

Like many features in top-quality digital avionics, BITE must be a

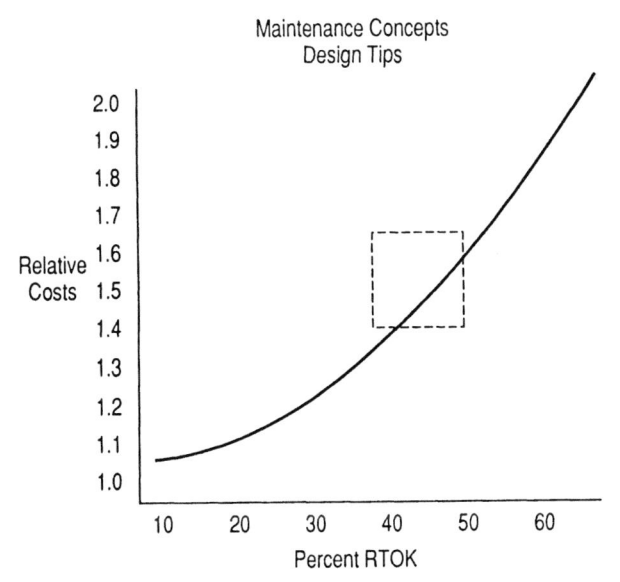

Figure 6.1 Effect of retest O.K. on life cycle cost. (*Curry, Ernest E., "STEP: A Tool for Estimating Avionics Life Cycle Costs," IEEE Aerospace and Electronic Systems Magazine, pp. 30–2, January 1989, vol. 4, no. 1.*)

part of the design from the beginning if it is to have maximum effectiveness. It cannot be added on after the design is nearly complete. The criticality and predicted failure rate of a unit drive the BITE design for that unit. BITE can be continuous, periodic, or on demand; the selection is driven by the criticality of the unit and its maintenance strategy. BITE can be based, in part, on a failure modes and effects analysis (FMEA) (see Chap. 9). BITE should use a building block approach, testing from the lowest-level function to the highest. It should test the inputs to the function before testing the function itself. BITE should be able to recognize and correctly identify at least 95 percent of possible faults in its host system and correctly isolate 80 to 90 percent of all faults to a single module, 90 to 95 percent to three or fewer modules and 95 to 100 percent to eight or fewer modules.[3]

BITE indicators should be clearly visible without maintenance personnel having to assume an uncomfortable or dangerous position. These indicators also should be able to indicate failure of the BITE itself. If BITE messages are displayed on a control display unit (CDU) on the aircraft, they should be in plain text; if fault codes are used, they should be mnemonically based. As onboard memory capacity increases and electronic library systems begin to be installed, the potential for electronic storage of the maintenance manuals on the aircraft becomes a very attractive option. If electronic display menus are used to present maintenance information, they should be designed based on the same human factors guidelines used in designing the flight menus.

BITE should be able to recognize faults in other LRUs as the cause of failure in its host LRU and to generate a message to that effect. Nuisance faults, such as power interrupts, should not be recorded except in the system generating the fault. In order to circumvent intermittent faults, multiple successive failures over an appropriate time period should be required before the BITE declares a unit failed.

BITE software should be solidly partitioned to permit changes, such as threshold levels, without having to revalidate all other software.

Any BITE intended for ground use only should be disabled when the aircraft brakes are released.

The amount of information available on a fault to maintenance personnel from BITE data should be a function of the maintenance level. Flight line maintenance needs only limited information, while shop maintenance requires more detailed information. As a goal, BITE should be capable of providing extensive data for engineering analysis in the case of a particularly vexing fault or problem in the LRU. A good example of the various levels of detail that may be available can be found in ARINC 629 (see Chap. 2).

MIL-STD-2165 Testability Program for Electronic Systems and Equipments establishes a series of tasks to be accomplished in the design of military avionics directed toward improving the testability and BITE of the avionics. Testability is defined as the extent to which a system or unit supports fault detection and fault isolation in a confident, timely, and cost-effective manner. One task calls for examining the compatibility between the BITE and the ATE to ensure agreement between the test configuration and acceptance criteria, as discussed earlier, and the structure of the item in terms of partitioning and accessibility. There are three facets of partitioning: physical, functional, and electrical. Another task calls for examining the effectiveness of the BITE and ATE: Are all potential faults covered by the BITE or ATE? MIL-STD-2165 requires for the BITE that 95 percent of all possible critical failures be detected in ≤ 1 s and that 100 percent be detected ≤ 1 min. Eighty-five percent of all other failures must be detected in ≤ 1 min.[4]

ARINC Report 604 Guidance for Design and Use of Built-in Test Equipment lists four types of tests that BITE can perform:

LRU power-up self-test: Automatically executed every time power is applied to confirm dispatchability of the LRU. Should include a central processing unit (CPU) self-check, memory sum checks, peripheral circuits, and input-output ports checks.

In-flight fault recording: Automatic passive monitoring and recording of LRU faults, not to be confused with self-checks performed as part of flight critical redundancy concepts.

LRU replacement verification test: Basically the same as the LRU power-up self-test with the possible addition of data reasonableness checks.

System performance test: This is a more complex test than the others and requires specific prerequisite conditions and interactive control actions. A test of this type checks more functions than a simple power-up test does and can be used for tasks such as checking alignment of sensors and actuators.

All of the above tests, except the last one, should be designed to be run by one person. The last test requires two or more persons so that one can observe the action or status of the hardware under test. BITE should exercise the hardware sufficiently to determine if it meets the performance requirements but should not drive it against mechanical stops.[5]

The CFDS builds on the capabilities of the BITE to consolidate information on failures from all of the avionics systems into a single database along with auxiliary information such as airspeed, altitude, time, and date when the failure occurred. The CFDS should be accessible from the flight deck via a CDU so the effect of operating circuit breakers can be observed during troubleshooting. However, many airlines believe that access to the CFDS should be disabled during flight to prevent the crew from being diverted from their primary task of flying the aircraft.

The A-320 CFDS, shown in Fig. 6.2 represents the state of the art for commercial transports. It can monitor up to 70 systems, but not all of them directly, and can store the 200 most recent faults or all faults for the last 63 flight legs. The CFDS also records Greenwich mean time (GMT), flight number and phase, and any cautions and/or warnings that were in effect at the time of the fault. The interface with the CFDS is through a multifunction control/display unit (MCDU) in the cockpit. All fault messages are displayed on the MCDU screen in standard formats using plain English. There are provisions for connecting the CFDS to the cockpit printer and to the aircraft communications addressing and reporting system (ACARS) (or equivalent) VHF data link. This latter provision allows advance notification to maintenance personnel at the flight destination who can begin preparations to work on the aircraft when it arrives.[6]

A particularly noteworthy feature of the A-320 CFDS is the assignment of each fault to one of three classes:

- *Class 1:* May affect flight in progress or dispatch for the next flight; displayed immediately to the flight crew.

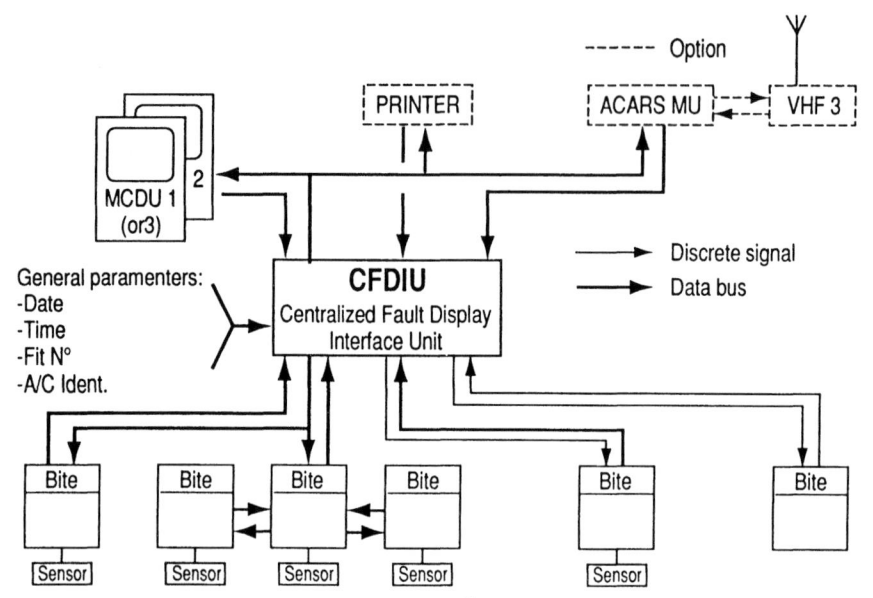

Figure 6.2 A-320 central fault display system.[6]

Maintainability and Reliability 111

- *Class 2:* Does not affect flight in progress or dispatch for next flight but requires timely rectification; displayed after landing or at any time upon request.
- *Class 3:* Deferred until routine maintenance; can be displayed at any time upon request.

Classification of the faults helps to ensure that they receive the proper attention.

Automatic Test Equipment Speeds Maintenance

Along with sound design and sophisticated built-in test equipment, the third ingredient of an easily maintained digital avionics system is ground-based ATE. Affordable, portable (for military applications), and versatile, ATE is necessary for the successful diagnosis and repair of LRUs, shop replaceable units (SRUs), and their subassemblies or components. ATE should be capable of the following functions:

- End-to-end testing of LRU, SRU, or other component
- Isolating faults to the component level
- Developing, debugging, and compiling software written in the ATLAS language (see the next section).

Well-designed ATE should offer a number of options to the operator including single-step testing, stop/no stop on failure, operator alert in case of failure, and test report print/no print. In addition, ATE state data, such as failure of the unit under test (UUT), name and step of the current test, loop count, and ATE failure, should be shown on an annunciator panel.

Military ATE should have transportability as its most important attribute. Rapid deployment requirements for aircraft dictate that the ATE should be small, lightweight, and rugged and should require minimum support. (The emergence of the two-level maintenance concept dictates that the BITE be expanded to eliminate the need for the avionics maintenance shop and associated ATE. Any avionics repairs would be performed at the depot.) Military ATE should also be fast to reduce maintenance time and therefore increase aircraft sortie generation rate.

The U.S. Air Force modular automatic test equipment (MATE) is an example of a modern set of military ATE. "MATE is a set of concepts and tools which can be applied to both the acquisition and the development of automatic testing support for any weapon system."[7] MATE achieves large ATE cost savings through such concepts as standardizing the ATE/operator and ATE/UUT interfaces. The use of MATE is growing, spurred in part by Air Force regulations which mandate its application to all new weapon systems acquisitions.

A cornerstone of MATE is the MATE control and support software (MCSS). MCSS is written in Jovial and operates on a 1750A processor. It contains the MATE operating system (MOS), MATE on-line editor (MOLE), MATE IEEE standard 716 C/ATLAS compiler (MAC), and the MATE test executive.[7]

For civil ATE there is a different set of design drivers. Civil ATE should be affordable and versatile since it is a capital investment. Versatility ensures that when a new generation of avionics comes along, the airlines will not have to purchase new ATE just for those avionics.

ATLAS: A Giant of a Test Language

Digital avionics and their inherent flexibility bring with them the potential (and need) for automatic testing. The airlines were one of the first groups to recognize the need for automated testing and consequently developed ATLAS in the late 1960s. This original version of ATLAS proved to be very popular and was used in many nonavionic applications. Because of this wider use, responsibility for maintaining and updating the language was transferred from ARINC to the Institute of Electrical and Electronics Engineers (IEEE). Today the senior ATLAS publication is ANSI/IEEE Standard 416 ATLAS Test Language.

A standard test language provides a common link between the avionics designer, test specification writer, test equipment designer, and the user. The special power of ATLAS lies in its ability to serve the dual roles of enabling the writing of a clear, concise test specification and then serving as the language to be used in the ATE.

ATLAS is a standard, abbreviated English language used in the preparation and documentation of test specifications which can be implemented either manually or with automatic or semiautomatic test equipment. The version of ATLAS discussed here is based on ARINC Specification 616 Avionics Subset of ATLAS Language and is in wide use in civil avionics. A similar version of ATLAS, IEEE Standard C/716 ATLAS is used by the military avionics community. In both cases, ANSI/IEEE Standard 416 is the parent document.

There are three levels of ATLAS: standard ATLAS as contained in the ANSI/IEEE standard, subset ATLAS which uses a limited number of terms from the standard ATLAS and uses them in strict conformance with the standard, and adapted ATLAS which closely conforms to the parent standard but potentially may have modified vocabularies and syntax distortions and is intended for very limited applications. As suggested by the title of ARINC 616, this section deals with Subset ATLAS (Avionics).

An ATLAS test specification generally contains four parts: the BEGIN program statement, an optional preamble structure, a main procedure structure, and an END program statement. The test specification is written for the UUT and is independent of any potential test equipment to ensure maximum transportability. The five elements of the ATLAS language, verbs, nouns, modifiers, connectors, and delimiters, are used in constructing the test specification.

The core of the test specification is the preamble and the procedural section. The preamble includes all the necessary information to configure the ATE for the proper execution of the procedural section. Typical information in the preamble includes:

Identification of the UUT

Listing of the independently executable modules and submodules in the procedural section by name and statement (line) number

Definition of test resource requirements, messages, functions, and procedures

Declaration of data storage requirements

Preload of data for use in test execution

The procedural section of the test specification contains the individual test procedures. This section should be organized as follows:

Identification of the UUT (repeat from the preamble)

Safety tests (optional)

Power supply tests

Readout of inflight failure storage memory

BITE tests

Functional tests

Functional tests are the most important part of the test specification. These tests should be divided into modules which correspond to the functions of the UUT. If needed, submodules can be matched to subfunctions or cards of the UUT. Modules should be independent and capable of being selected on an individual basis as needs dictate and should be designed to require minimum operator intervention. If intervention is needed, it should be either early or late in the test to increase the productivity of the maintenance personnel.

The test specification should be flexible and should allow the operator to choose from a large menu of options including selection and/or omission of individual tests, step-by-step execution of a given test, test repetition, and print out of test results.

Reliability Is Equal Parts of Art and Science

Military avionics designers have always focused extraordinary amounts of attention and resources toward achieving ever higher levels of reliability. Reliability is also important to civil avionics designers, especially in the case of flight critical avionics; however, civil designers often must balance reliability against other design goals such as cost. (Chapter 11 discusses the trade-off of reliability versus life cycle cost.)

MIL-STD-721C defines reliability as:[8]

(1) the duration or probability of failure-free performance under stated conditions, or (2) the probability that an item can perform its intended function for a specified interval under stated conditions. (For nonredundant items this is equivalent to definition (1). For redundant items this is equivalent to the definition of mission reliability.)

Like maintainability, reliability must be factored in from the beginning of the design process, and it must consider the mission of the aircraft. Mission requirements such as 30-day surge operations without maintenance will clearly have an impact on reliability goals.

The traditional approach to reliability analysis has been to use failure prediction methods as found in many well-known reliability handbooks. The best known of these handbooks is *MIL-HDBK-217 Reliability Prediction of Electronic Equipment*.[9] This handbook contains classical reliability prediction methods that begin with such input data as types of parts and number of each, the quality level of the part, and its duty cycle and operating environment. These data are then used in probability of failure equations such as the one shown below for a single string, series function:[9]

$$\lambda_{Equip} = \sum_{i=1}^{n} N_i \lambda_{Gi} \pi_{Qi}$$

where λ_{Equip} = total equipment failure rate (failures/10^6 h)
  λ_{Gi} = generic failure rate for ith generic part type (failures/10^6 h)
  π_{Qi} = quality factor for ith generic part type
  N_i = quantity of ith generic part type
  n = number of different generic part types

More sophisticated versions of this equation recognize the effects of the avionics operating environment with special attention being given to temperature. Table 6.1 summarizes the strong effect of the environment on the mean time between failures (MTBF). For example, if a

TABLE 6.1 Part Environment Conversion Factors (MTBF Multiplier)[10]

To environment class	From environment class		
	Space	Airborne Inhabited Cargo	Airborne Uninhabited Fighter
	Multiply the MTBF by		
Space	x	4	14
Airborne inhabited cargo	0.25	x	4
Airborne uninhabited fighter	0.07	0.25	x

space-rated part is operated in an uninhabited fighter aircraft, its MTBF will be only 7 percent of its space MTBF value.

In a similar vein, operating a part at other than room temperature will have an effect on its MTBF. Lowering the temperature of the part 10°C raises the MTBF 25 percent (1.25). Likewise, raising the temperature of the part 10°C lowers the MTBF 20 percent (0.80).[3]

Derating (electrical, thermal, and/or mechanical) is a common design practice to reduce the stresses on component parts. That is, a part is operated at conditions more favorable than those for which it was designed. Typical recommended electrical deratings are shown below:[10]

VHSIC chips*	Supply voltage noise, ±3%	Frequency, 75%
	Output current, 70%	Maximum T_j, 85°C
Connectors	Voltage, 50%	Current, 50%
LEDs	Forward current, 50%	Maximum T_j, 95°C
Fiber optic cables	Bend radius, 200%	Fiber tension, 20%

*Very high-speed integrated circuit.

All electronic parts used in military avionics design must be (1a) on the Program Parts Selection List or (1b) selected in accordance with MIL-E-5400 and MIL-STD-454 and (2) approved by the customer. Semiconductors must be selected in accordance with MIL-STD-454 Requirement 30. One notable element of Requirement 30 is that all semiconductor devices must be hermetically sealed in glass, metal, metal oxide, ceramic, or combinations of these. Microcircuits must be selected in accordance with MIL-STD-454 Requirement 64. Two notable elements of Requirement 64 are (1) fault coverage of devices must be determined (built-in test should detect 98 percent of all possible faults on the chip) and (2) devices must be selected on the basis of overall life cycle costs.[10,11]

The traditional approach to reliability prediction described above has had, in the judgment of some persons, only limited success. "The fundamental concept of a constant failure rate, accelerated by stresses, does not accurately describe current electronic technology."[12] In many cases the predicted reliability of an LRU determined using this traditional approach is far below (worse than) the actual reliability achieved in service. This large difference between predicted and actual reliability has long been recognized, and in recent years it has begun to receive some attention.

A partial explanation is that MIL-HDBK-217 predicted MTBFs are conservative. Actual field experience is often substantially better. For example, in the electronic display systems on the B-757 and B-767 component failure rates were 20 percent of MIL-HDBK-217D predictions. Possible explanations are that the predicted failure rates were unduly pessimistic and the favorable environmental effects were not fully recognized. The electronic flight instrument system (EFIS) in-service reliability is 200 percent of prediction, and the engine indicating and crew alerting system (EICAS) reliability is 150 percent of prediction.[13]

According to Wong,[14]

> The most important reason for the inaccuracy lies in the fact that many critical factors that influence reliability in the first order manner are not even considered in the existing models.... The models are often simple approximations without scientific basis.

Leonard offers a similar observation:[12] "Part failures are generally due to design errors, workmanship, supplier inadequacies, or similar human-related events...." He believes there is too much emphasis on component part failure rate and a failure to recognize the contribution of "inadequacies in design and workmanship" in traditional failure prediction methodologies.[12] He cites as evidence experience with a civil transport flight management computer (FMC). The FMC contained 4000 electronic components, weighed 40 pounds (lbs), and consumed 180 W of power. For 60 percent of all removals, the FMC retested O.K.; no fault was found and the unit was returned to service. Of the 40 percent that were faulty, 80 percent of the failures were due to workmanship and handling, that is, failed solder joints and printed circuit boards, bent pins, and so forth. Therefore, only 8 percent of all FMC removals were due to parts failures. Of this 8 percent, 80 percent of the failures resulted from the parts being improperly applied or designed. Therefore, less than 2 percent of all FMC removals were due to true electronic failures.[15]

While the *Reliability Engineer's Handbook* recognizes the direct effect of temperature on failure rate, there is strong evidence to suggest that "In general, for reasonable temperature extremes, thermal cycling effects outweigh sustained temperature effects."[14] Failure rate show a good correlation with the number of thermal cycles an LRU has experienced. Furthermore, there is also evidence to suggest an exponentially decaying failure rate for LRUs based on the time since being powered up. Shurman has shown for many types of military and civil aircraft with a wide range of typical sortie times that the LRU failure rate decreases with sortie elapsed time.[16] This short-term reliability trend may support the maxim: "If the avionics work to the end of the runway, they'll work to the end of the mission."

References

1. MIL-STD-470A Maintainability Program for Systems and Equipment, Notice 1, 26 August 1987.
2. MIL-STD-721C Definition of Terms for Reliability and Maintainability, 12 June 1981.
3. *RADC Reliability Engineer's Toolkit,* Rome Air Development Center, 1988. Available from Reliability Analysis Center, P.O. Box 4700, Rome, NY 13440-8200.
4. MIL-STD-2165 Testability Program for Electronic Systems and Equipments, 26 January 1985.
5. ARINC Report 604 Guidance for Design and Use of Built-In Test Equipment.
6. Vauversin, F., and Potocki de Montalk, J. P., "Central Fault Display System," *XVI International Congress on Aeronautical Sciences,* Jerusalem, Israel, ICAS-88-2.8.3, 1988.
7. *MATE Introduction and Users Notebook,* U.S. Air Force, May 1989.
8. MIL-STD-721C Definition of Terms for Reliability and Maintainability, 12 June 1981.
9. MIL-HDBK-217E Reliability Prediction of Electronic Equipment, 27 October 1986.
10. MIL-E-5400T Electronic Equipment, Airborne, General Specification for, Amendment 3, 14 May 1990.
11. MIL-STD-454M Electronic Equipment, Standard General Requirements for, Notice 2, 3 June 1991.
12. Leonard, C. T., and Pecht, Michael, "How Failure Prediction Methodology Affects Electronic Equipment Design," *Quality and Reliability Engineering International,* vol. 6, pp. 243–9, 1990.
13. Moorhead, Michael E., "Implications of New Aircraft Avionics Reliability Performance," *Proceedings of 30th Annual Technical Meeting,* Institute of Environmental Sciences, pp. 232–5, 1984.
14. Wong, Kam L., "What Is Wrong with the Existing Reliability Prediction Methods?" *Quality and Reliability Engineering International,* vol. 6, pp. 251–7, 1990.
15. Leonard, C. T., "Mechanical Engineering Issues for Electronics," *Proceedings of 1989 IEEE National Aerospace and Electronics Conference (NAECON),* 89CH2759-0, pp. 2043–9, 1989.
16. Shurman, M. B., "Time-Dependent Failure Rates for Jet Aircraft," *Proceedings of the IEEE 1978 Annual Reliability and Maintainability Symposium,* 78CH1308-6, pp. 198–203, 1978.

Bibliography

Leonard, C. T., and Pecht, M., "Failure Prediction Methodology Calculations Can Mislead: Use Them Wisely, Not Blindly," *Proceedings 1989 IEEE National Aerospace and Electronics Conference (NAECON),* 89CH2759-0, pp. 1887–92, 1989.

Chapter 7

Architectures

The emphasis in earlier chapters has been on defining the requirements for the avionics system and introducing the building blocks for it. This chapter reviews fundamentals of design and philosophy for avionics systems architectures and presents several examples.

Fundamentals of Architectures

Selecting an avionics system architecture is the greatest challenge facing the designer. The architecture must conform to the aircraft and mission requirements and the avionics system requirements derived therefrom, while successfully meeting the hardware and software assessments and validations described in Chaps. 9 and 10, respectively.

Modern, large-scale integrated electronics along with data buses have enabled the designer to choose from three fundamental types of architecture: centralized, federated, and distributed. The optimum architecture is established only by an iterative process of design and assessment.

A centralized architecture is characterized by signal conditioning and computations taking place in one or more computers in a line replaceable unit (LRU) located in the avionics bay with sensor and command signals transmitted over data buses. This has been the traditional architecture as digital avionics evolved from its analog ancestors. A centralized architecture has the following advantages: (1) all computers are located in a readily accessible avionics bay, (2) the environment for the computers is relatively benign, which simplifies equipment qualification and certification, and (3) the software is more easily written and validated since there are only a few processor types and a few large programs that can be physically integrated. The disadvantages of a centralized architecture include: (1) many long buses to collect and distribute data and commands and (2) increased vulnerability to damage from a single hazardous event if it were to occur in or near the avionics bay. The centralized design also makes software changes somewhat difficult since the possible impact of the changes on a relatively large number of blocks of codes in the total program has to be examined. Partitioning, as discussed later in this section, is difficult.

A federated architecture is characterized by each major system, such as thrust management or flight management, sharing input and sensor data from a common set of hardware, and consequently sharing their computed results over data buses. This type of architecture permits the independent design, configuration, and optimization of the major systems while ensuring that they are using a common database. Partitioning is an intrinsic feature of this architecture. Changes in system hardware or software are relatively easy to make. Federated architectures are commonly found in commercial transport aircraft designed and built in the 1980s.

A distributed architecture has multiple processors throughout the aircraft that are assigned computing and control tasks in real time by executive software as a function of mission phase and/or system status. Limited signal processing may be performed in or near the sensors or actuators. The advantages of a distributed architecture include fewer, shorter buses, faster program execution, intrinsic partitioning, and somewhat reduced vulnerability since a hazardous event will not destroy a substantial fraction of the total system capability. The disadvantages include a potentially greater diversity in processor types, which complicates software generation and validation, and spares stocking. Each processor contains many lines of code that perform a variety of functions. Since this software is identical in many processors, some economy of scale is achieved, but this savings may be offset by the need to make subsequent software changes in many processors. Furthermore, some of the processors may be in more severe, less accessible environments, such as wings and empennages. Distributed architectures are being examined for advanced avionics systems.

Partitioning, or *brickwalling,* is an architecture feature that limits a failure to the subsystem in which it occurred. In addition, the physical and operational effects of the failure are not allowed to propagate to the rest of the system. Examples of failures which are particularly amenable to partitioning include a short in a power lead and a bit hang up in a processor. Partitioning can be between peer units and/or between levels in a hierarchical system. Partitioning is an especially effective design technique for achieving the performance goals required by FAR 25.1309.

Examples of Modern Architectures

This section presents examples of modern military and civil aircraft avionics architectures. These architectures are given as examples only to show the range of possible options and to guide the designer in developing an architecture that best meets the need at hand.

Figure 7.1a shows the digital avionics architecture for the U.S. Air Force F-16 C/D fighter aircraft. The cornerstones of this highly integrated architecture are the four dual redundant MIL-STD-1553 data buses. For three of the buses, AMUX, BMUX, and DMUX, the primary bus controller (BC) function is provided by the fire control computer. The stores management system central interface unit provides the primary BC function for the WMUX bus. Each bus generally supports a top-level function of the system (e.g., the AMUX supports navigation and engine control, the BMUX supports defensive avionics, the DMUX handles displays, and the WMUX supports the remote interface units).

The digital flight control system (DFCS) is perhaps the most impor-

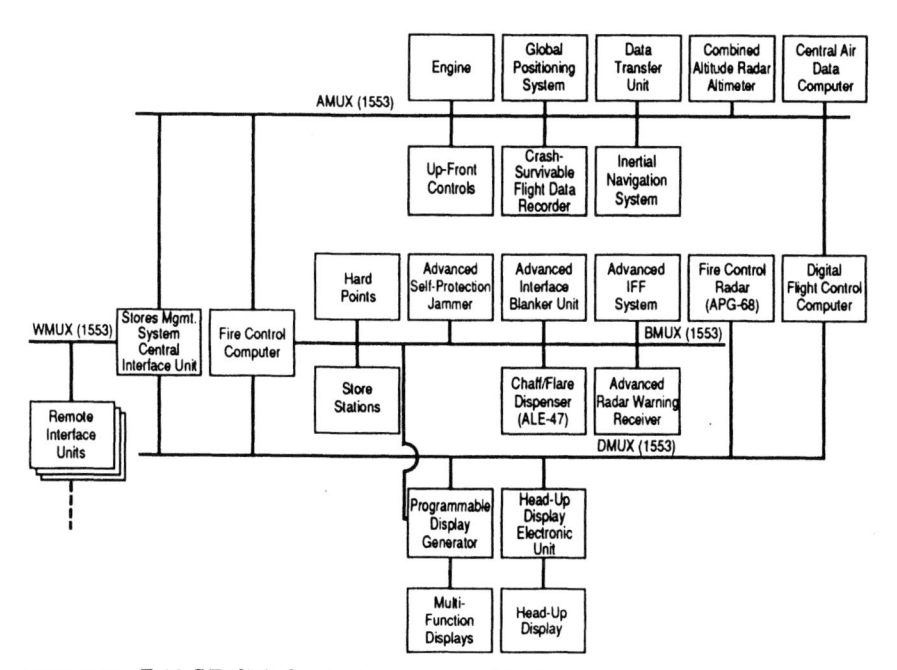

Figure 7.1a F-16 C/D digital avionics system architecture.

tant system on the aircraft. Because of its criticality, it is designed for a probability of failure resulting in the loss of the aircraft of less than 10^{-7} per flight hour and the probability of a mission abort due to DFCS failure of less than 10^{-5} per flight hour. In order to achieve these very low probabilities, the DFCS has quadruple redundancy. Figure 7.1b shows one of the four identical flight control computer (FLCC) channels.[1]

Each FLCC has a computation section and a communication section that are linked by a common scratch pad memory (XPAD Mem. in Fig. 7.1b). The computation section includes the central processing unit (CPU), 48K of program memory, a dedicated scratch pad memory, and a 256-word failure record memory. These three memories can be accessed only by the CPU. The program and failure record memories are nonvolatile. The failure record memory stores data on in-flight failures for post-flight retrieval by maintenance personnel. The CPU is a single chip microprocessor that executes the MIL-STD-1750A Instruction Set Architecture. All software is written in Jovial.[1] The communication section of each FLCC interfaces with the other FLCCs, the MIL-STD-1553 data buses, analog to digital (A/D) and digital to analog (D/A) converters, discrete registers, and output latches.[1]

Pave Pillar, shown in Fig. 7.2, is a generic, conceptual architecture developed by the U.S. Air Force "specifically targeted for advanced tactical fighters, and in general for all military aircraft applications.... either air-to-air or air-to-ground missions."[2]

Even though Pave Pillar is a generic architecture, the design process was initiated like an actual architecture by defining the avionics system major performance requirements listed below:

- Mean time between critical failures:* 70 h
- Mean time to repair (critical functions): 1.25 h (includes troubleshooting: 10 min)
- Fault detection: 99 percent of all possible faults
- Fault isolation: 98 percent of all possible faults
- Two-level maintenance†
- Sortie rate: ≥4.5 per day
- Abort rate: ≤1 percent
- Combat turnaround time: ≤15 min

---

*A critical failure is a failure of an essential system that would prevent the successful completion of the mission.
†Two-level maintenance means that repairs are either made on the aircraft or in the depot. There is no intermediate-level maintenance.

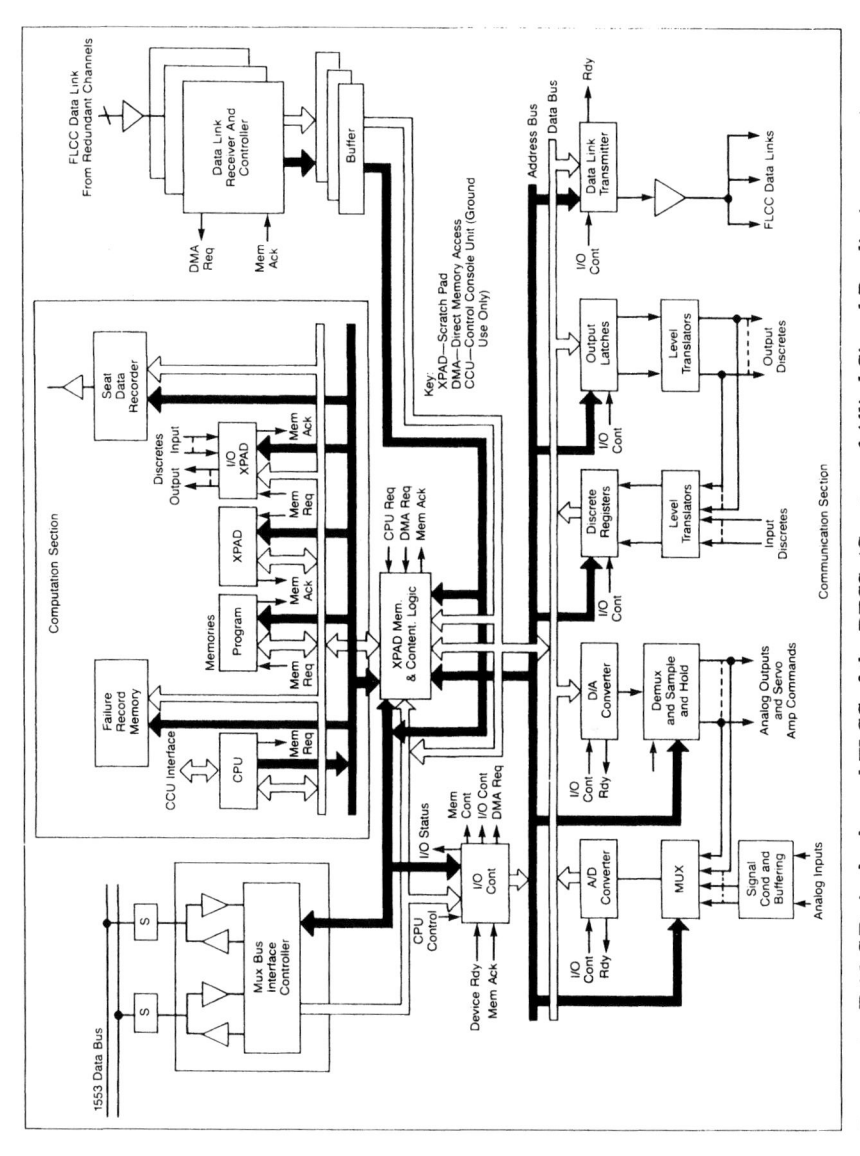

Figure 7.1b F-16 C/D single-channel FLCC of the DFCS. (*Courtesy of Allied Signal Bendix Aerospace.*)

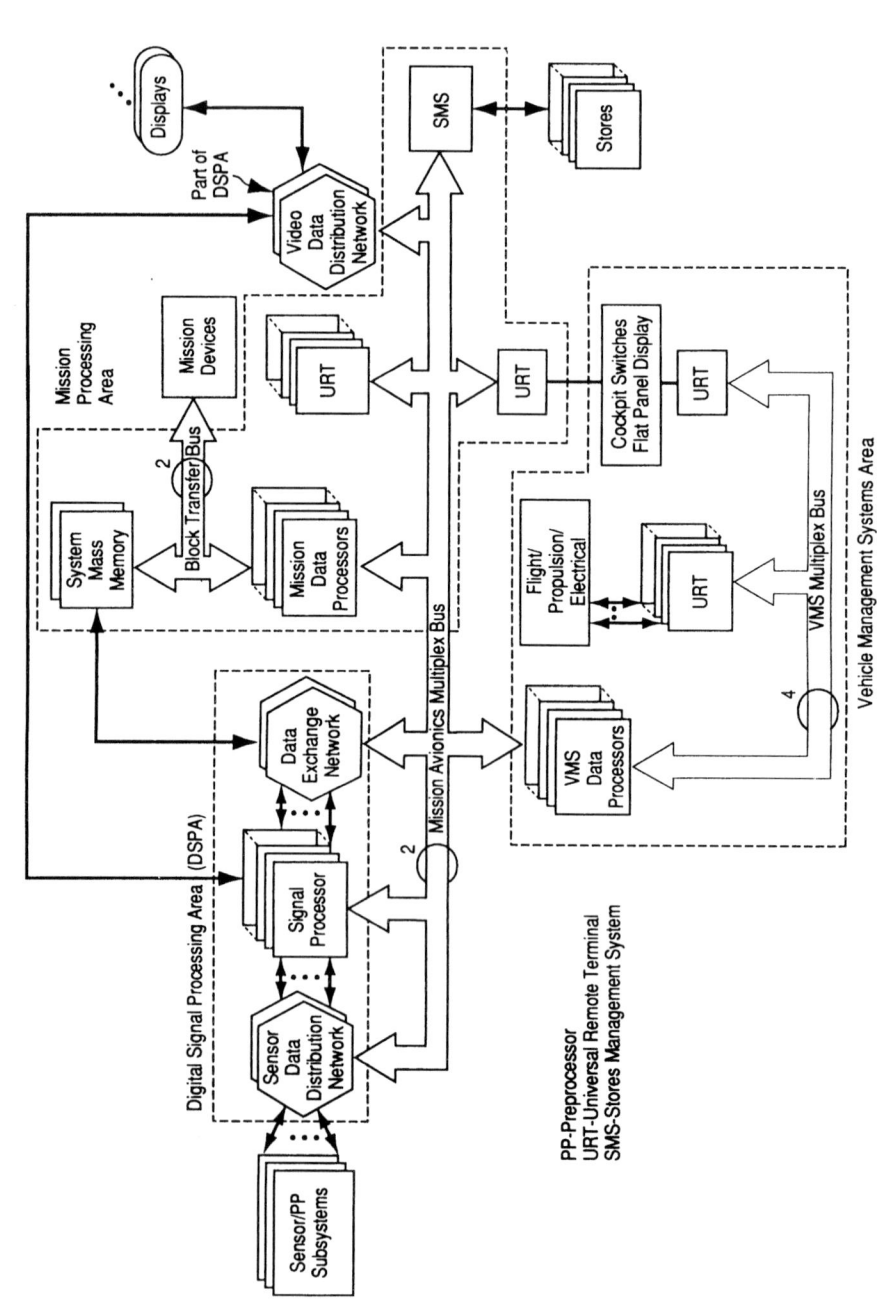

Figure 7.2 Pave Pillar conceptual architecture.[2]

Pave Pillar is divided into four functional areas:

- Digital signal processing
- Mission processing
- Vehicle management systems
- Avionics systems control

The digital signal processing functional area routes data from the receiving sensor to and among selected signal processing elements and processes radar, electronic warfare, image, and communications/navigation/identification signals.

The mission processing functional area uses its resident system executive software to manage fire control, target acquisition, navigation, stores, and the crew station configuration. It manages and reconfigures, as necessary, the digital signal processing function described above. The mission processing function also collects health and status information on all of the avionics.

The vehicle management system (VMS) handles all "inner loop" (in control parlance) control functions such as flight control, propulsion (including inlet and vector thrust) control, air data measurement, aircraft inertial measurement, and electrical power control. The VMS also manages the utility systems including fuel measurement, transfer, and inerting; hydraulics; environmental measurement and control; life support; crew escape system; and the landing gear.* Because of the flight critical nature of the VMS functions, the VMS multiplex bus is a quadruple redundant high-speed data bus (HSDB; see Chap. 2).

Avionics systems control is all of the software in the system. Figure 7.3 shows how the system software is partitioned. The system executive software is resident in all of the mission data processors (MDPs), and the MDP that gains control during power up is the system executive during that operating period or until it fails. Additional executive functions and the kernel executive are resident in all of the other MDPs. The mission management and mission applications software can execute in any MDP.

The three hardware functional areas are connected by the dual redundant mission avionics multiplex bus, a high-speed data bus as described in Chap. 2.

Because of the flight critical nature of the functions performed by the avionics, fault tolerance is required in both hardware and soft-

---

*The VMS concept originated by Pave Pillar has become widely recognized and is being applied to many new aircraft, not just advanced tactical aircraft.

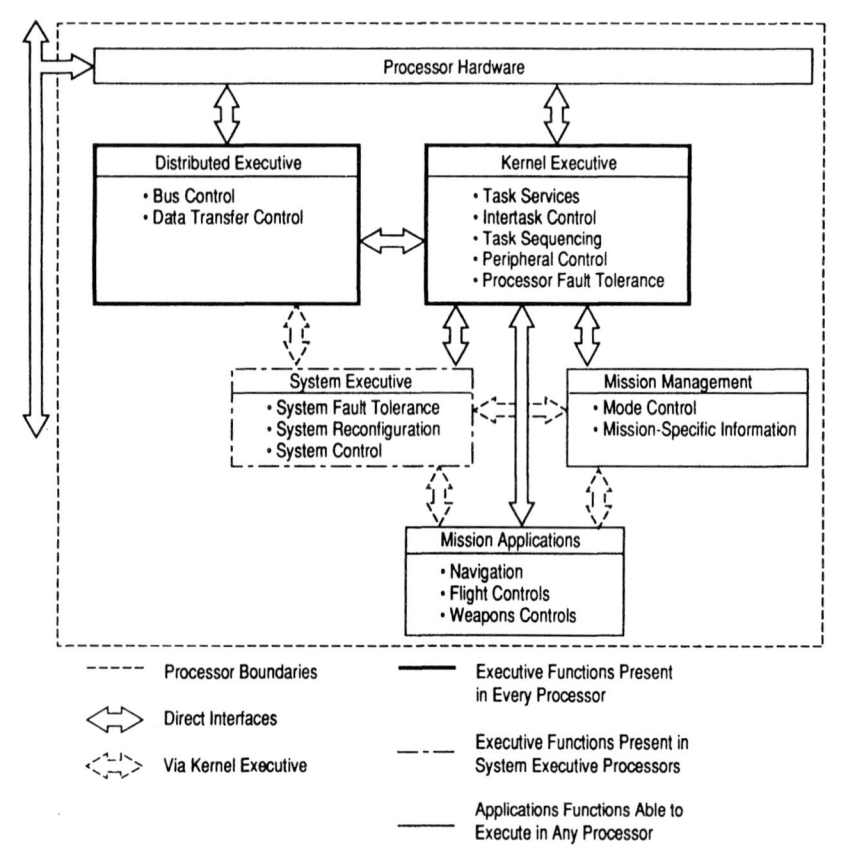

Figure 7.3 Partitioning of the Pave Pillar system software.[2]

ware. The fault-tolerant design must detect and correct (or compensate for) soft errors (memory bit or transmission errors) and detect, locate, and isolate, to the line replaceable module- (LRM) level, nonautomatically correctable errors so that graceful reconfiguration can be accomplished. Reconfiguration must be within the given system (e.g., digital signal processing, MDP, or VMS).

Pave Pillar is a modular avionics architecture with 23 types of LRMs primarily based on MIL-STD-1750 processors and advanced development models as listed in Table 7.1.

Pave Pillar LRM physical characteristics are modeled after those found in MIL-M-28787 Modules, Standard Electronic, General Specification for, discussed in Chap. 8. All electrical components must meet MIL-M-38510 Microcircuit, General Specification for, discussed in Chap. 6. Each LRM must be directly accessible for maintenance (i.e.,

TABLE 7.1 Pave Pillar Modules[2]

Part A: Modules based on MIL-STD-1750A and advanced development model technology	
VHSIC 1750A processor	Avionics bus interface
Bulk memory/nonvolatile bulk memory	MIL-STD-1553B I/O
Dedicated interface	Floating point processing element
Global memory	Data network element
Sensor interface	Element supervisor unit
Timing and control generator	I/O terminator
DC-DC 5-V power converter	DC-DC multivoltage power converter
Power conditioner	

Part B: Modules based on other than MIL-STD-1750A and advanced development model technology
Fixed point vector processing element
Electronic warfare-oriented processing element
Biphase correlation processing element
MIL-STD-1760 stores I/O
Video
Network switch module
Display
Key generator

it must not be necessary to remove an LRM to perform maintenance on another LRM). The enclosure (cabinet) for the LRMs must meet MIL-E-5400 Electronic Equipment, Airborne, General Specification for, and MIL-E-6051 Electromagnetic Compatibility Requirements, Systems.

Another example of a modern avionics architecture that reflects current thinking is that of the EuroFighter Aircraft (EFA). The EFA is a multinational fighter developed by England, Germany, Spain, and Italy to satisfy their respective needs for a twenty-first-century fighter aircraft. It has a flight critical fly-by-wire control system built by GEC Avionics and a utilities control system (UCS) built by Smiths Industries. The UCS, shown in Fig. 7.4, comprises five computers integrated by a MIL-STD-1553 data bus, and it provides automatic centralized control for the secondary power (electrical), hydraulic, fuel management, fuel gauging, environmental, and life support and crew sys-

128   Chapter Seven

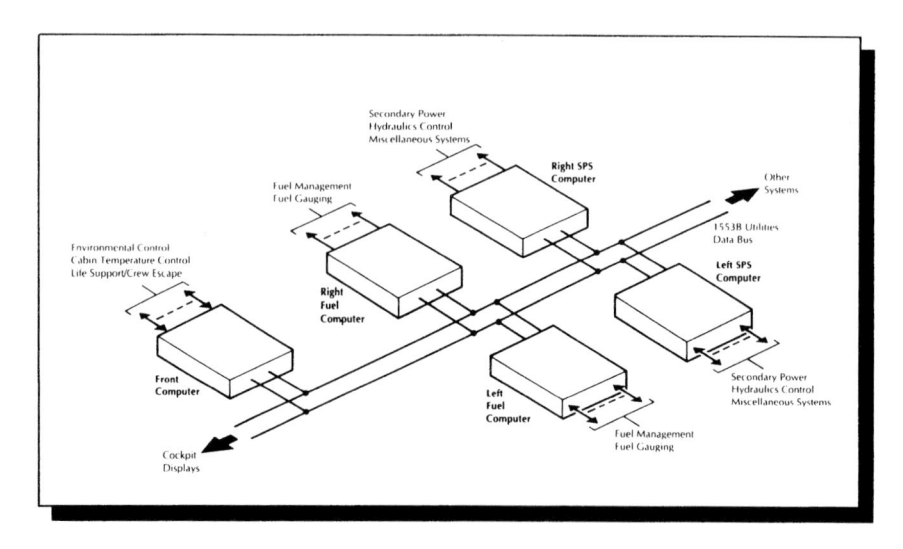

EFA Utility Control System

Comprises:	• Front Computer (CASA*) • Left Fuel Computer (SI+) • Right Fuel Computer (SI) • Left SPS Computer (Alenia) • Right SPS Computer (Alenia)	System Characteristics:	• Motorola MC68020 Microprocessor • MIL-STD-1553B Data Bus • Ada Software • Extensive System Monitoring • Extensive System and LRU BIT • Interface with Digital Cockpit • Solid State Switching of Aircraft Loads • Custom-Design ASIC Technology
Functions:	• Environmental Control • Cabin Temperature Control • Life Support/Crew Escape Monitoring • Fuel Management • Fuel Gauging • Secondary Power Control • Hydraulics Control • Miscellaneous Systems		*: Construcciones Aeronauticas, S.A. +: Smiths Industries

Figure 7.4   EuroFighter Aircraft utilities control system. (*Courtesy of Smiths Industries.*)

tems. These systems traditionally have been managed manually on an individual basis.* Note the dual redundant buses and computers, except a single front computer, reflecting the essential functions performed by the UCS. The UCS uses an MC 68020 microprocessor and the software is written in Ada.

In civil transport aircraft an architecture that promises to be a benchmark in the future is the Boeing B-777 Airplane Information Management System (AIMS). AIMS, with its partial, but significant, integration of hardware and of software, is a major step toward a completely integrated avionics architecture that promises enhanced per-

---

*The UCS is analogous to the Pave Pillar vehicle management system (VMS), although an important difference is that the VMS also performs the flight control function.

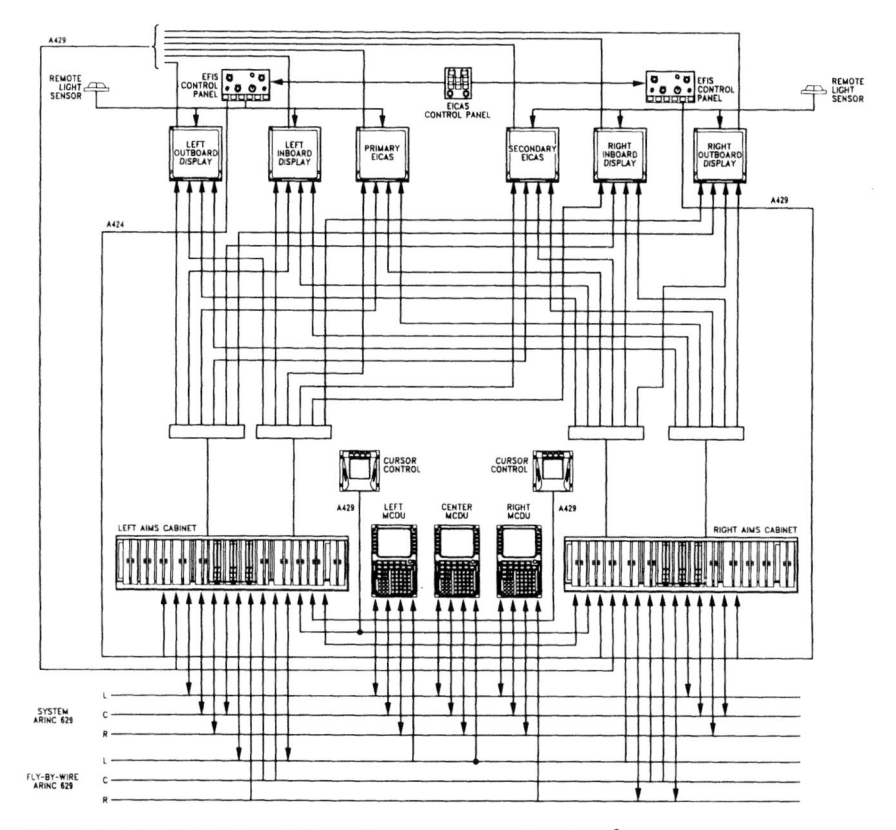

Figure 7.5 B-777 Airplane information management system.[3]

formance and reduced life cycle costs. The central features of AIMS, shown in Fig. 7.5, are the bread-box-size cabinets containing LRMs.[3]

AIMS functions performed in both cabinets include flight management, electronic flight instrument system (EFIS) and engine indicating and crew alerting system (EICAS) displays management, central maintenance, airplane condition monitoring, communications management, data conversion and gateway (ARINC 429 and ARINC 629), and engine data interface. AIMS does not control the engines or operate any internal voice or external voice or data link communications hardware. Subsequent generations of AIMS may include some of these latter functions.[3]

Within each cabinet the LRMs are interconnected by dual ARINC 659 backplane data buses. The cabinets are connected to the triplex redundant ARINC 629 system and fly-by-wire data buses and are also connected via the system buses to the three multifunction control display units (MCDUs) used by flight crews to interact with AIMS. The cabinets transmit merged and processed data over quadruple redundant ARINC 429 buses to the EFIS EICAS displays.[3]

Figure 7.6 shows the concept of common resources among multiple cabinets, a key feature of AIMS. Every cabinet has common design hardware consisting of power supplies, processors, memories, chassis, and shared input-output (I/O). In a given cabinet the only unique hardware is the I/O needed to interface with external equipment unique to the cabinet. The same pattern of commonality exists in software. Every cabinet has common design software for the operating system, utility (I/O, etc.), and hardware built-in test. In a given cabinet the only noncommon software is that required for the functions (and their associated built-in test) that are unique to the cabinet. Common hardware and software reduces development and manufacturing costs and the number of types of spare LRMs required and also expedites certification of the system. Maintenance costs are also reduced.[3]

Figure 7.7 shows the AIMS cabinet internal architecture. There are presently 11 LRMs, although the cabinet has a capacity of 13 LRMs to allow for growth as functions are added during the life of the aircraft. There are two power LRMs (each connected to the two power buses), four I/O LRMs, an autothrottle LRM, a modem LRM, and three central processor module (CPM) LRMs. The distribution of AIMS computational tasks among the CPM LRMs is also shown.[3]

In the AIMS architecture the high degree of function integration re-

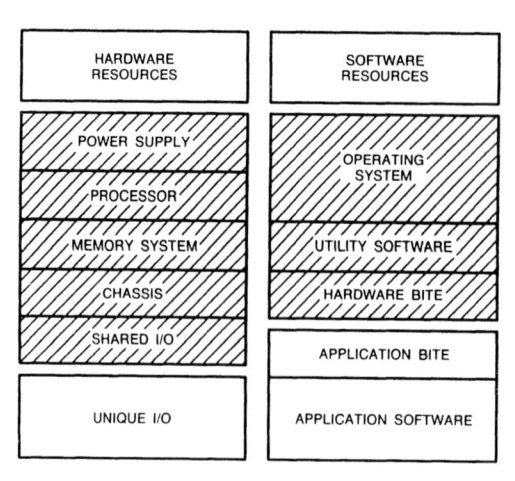

Figure 7.6 B-777 AIMS common resources.[3]

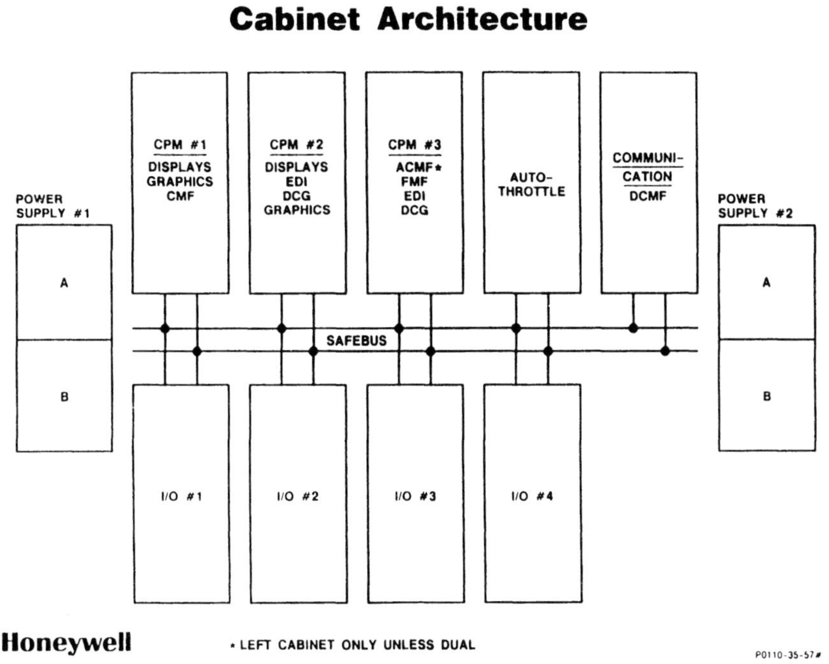

Cabinet Architecture

Figure 7.7 B-777 AIMS cabinet internal architecture.[3]

quires levels of system availability and integrity not found in traditional distributed, federated architectures. These extraordinary levels of availability and integrity are achieved by the extensive use of fault-tolerant hardware and software concepts. Fault tolerance and its associated high-integrity monitoring also means improved maintenance diagnostics and promises to reduce the chronic problem of unconfirmed removals and low mean time between unconfirmed removals (MTBUR).* Fault tolerance also means a very high probability of dispatch. Figure 7.8 shows the probability of aircraft dispatch as a function of days since maintenance action and how substantially the probability of dispatch exceeds the B-777 requirements.[3,4]

The Airbus Industrie A-320 commercial transport is the first aircraft of its type with a flight critical, fly-by-wire flight control system. There is a direct mechanical link to the rudder, but under normal flight conditions the rudder is controlled by the fly-by-wire control

---

*Unconfirmed removals are equipment removals from service in which subsequent diagnosis and troubleshooting of the equipment reveal no problems and the equipment successfully passes bench testing without any maintenance or repair actions. The equipment is then placed in the spares pool to await later use.

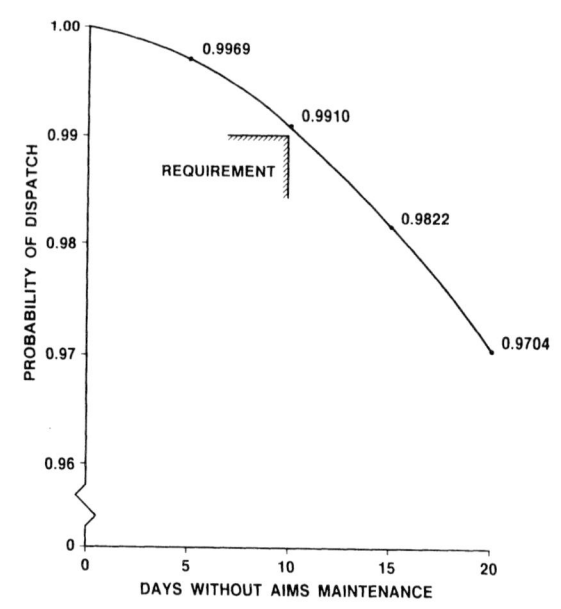

Figure 7.8 B-777 Probability of dispatch versus days without AIMS maintenance.[3]

system. There is also a direct mechanical link to the trimmable horizontal stabilizer.[5]

Seven computers form the heart of the flight control system:

- Two elevator aileron computers (ELACs)
- Three spoiler elevator computers (SECs)
- Two flight augmentation computers (FACs)

The allocation of these computers to the flight control surfaces is shown in Fig. 7.9. This architecture ensures that the failure of any one ELAC, SEC, or FAC will not substantially affect the performance of the flight control system.

The ELAC contains a pair of MC 68000 processors, each of which has approximately 40,000 words of software. One MC 68000 is the command unit and the other is the monitor unit to check the performance of the command unit. Dissimilar software is used in the pair of processors as a fault-tolerance technique. Each ELAC fits in a standard ARINC 600 6 modular concept unit (MCU) LRU.[5]

The SEC contains a pair of Intel 80186 processors, each of which has approximately 40,000 words of software. One 80186 is the command unit and the other is the monitor unit to check the performance of the command unit. Dissimilar software is used in the pair of processors as

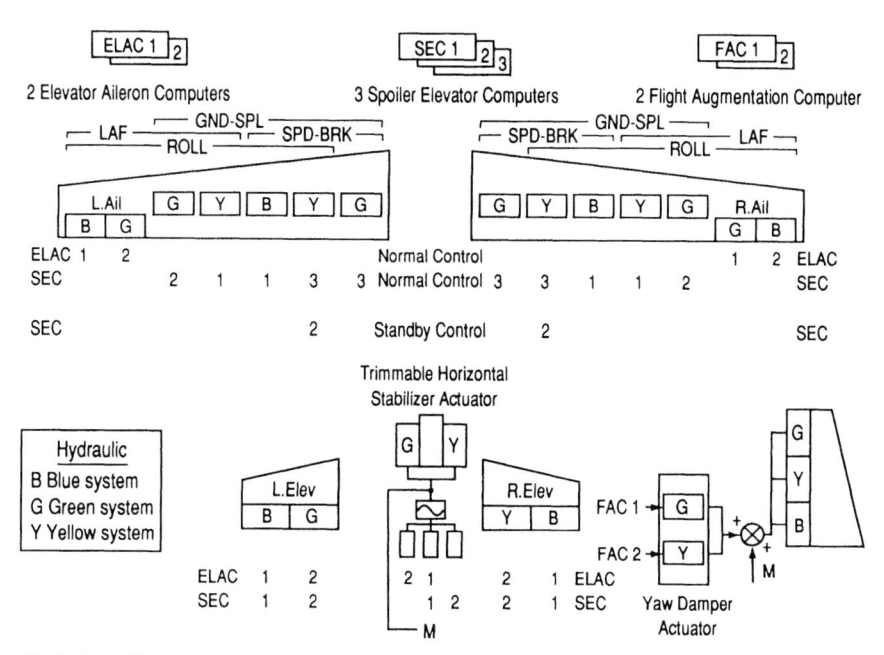

Figure 7.9 A-320 Electronic flight control system architecture. (*Courtesy of Airbus Industrie.*)

a fault-tolerance technique. Each SEC fits in a standard ARINC 600 8 MCU LRU.[5]

The functions of the ELACs and the SECs are identical except that ELAC control of the spoilers is via the SEC. The basic control law and all protection systems are in both the ELACs and the SECs. In case of ELAC failure, the ailerons are no longer active and the aircraft is controlled by the SECs. The ELACs and the SECs are designed and manufactured by separate divisions of Sextant Avionique to ensure maximum dissimilarity (see Fig. 7.10).[6]

The FAC has the same dual structure as the ELACs and SECs. Each contains 120,000 words of software. The FACs are also manufactured by Sextant Avionique.[5]

The Beechcraft Starship architecture is one of the most interesting and instructive contemporary ones. It is the fortuitous result of the confrontation between the severe constraints on power, weight, and volume always found in business jets and the proven capabilities of digital processors, displays, and data buses when used in an avionics application. The subsystem elements of the Starship avionics system

134 Chapter Seven

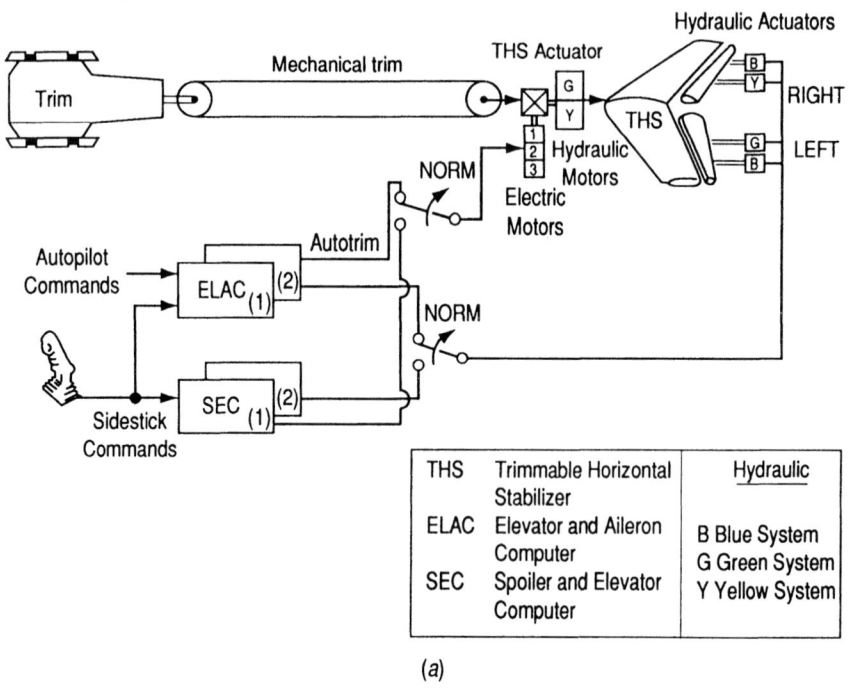

Figure 7.10a A-320 Pitch control system. (*Courtesy of Airbus Industrie.*)

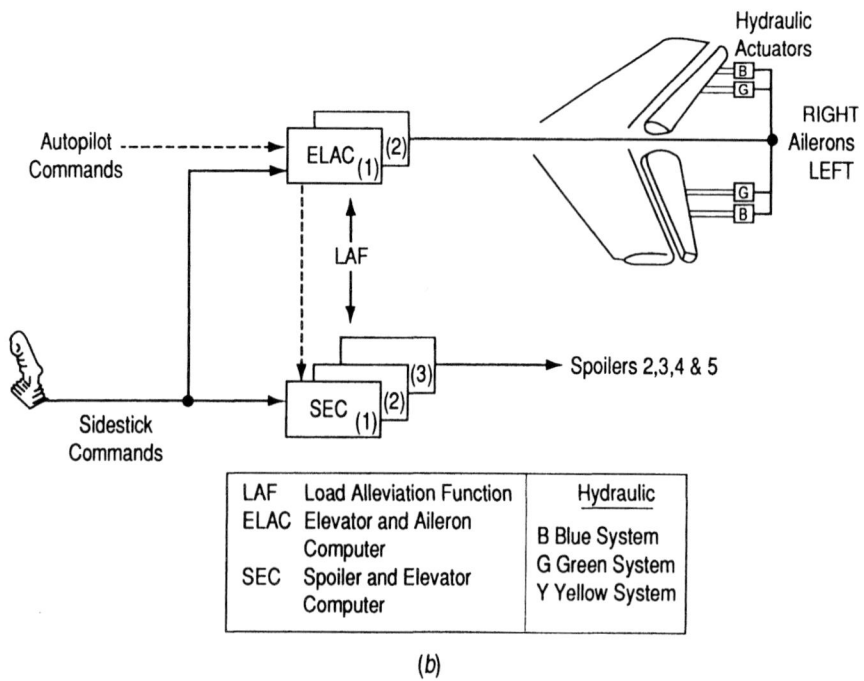

Figure 7.10b A-320 Roll control system. (*Courtesy of Airbus Industrie.*)

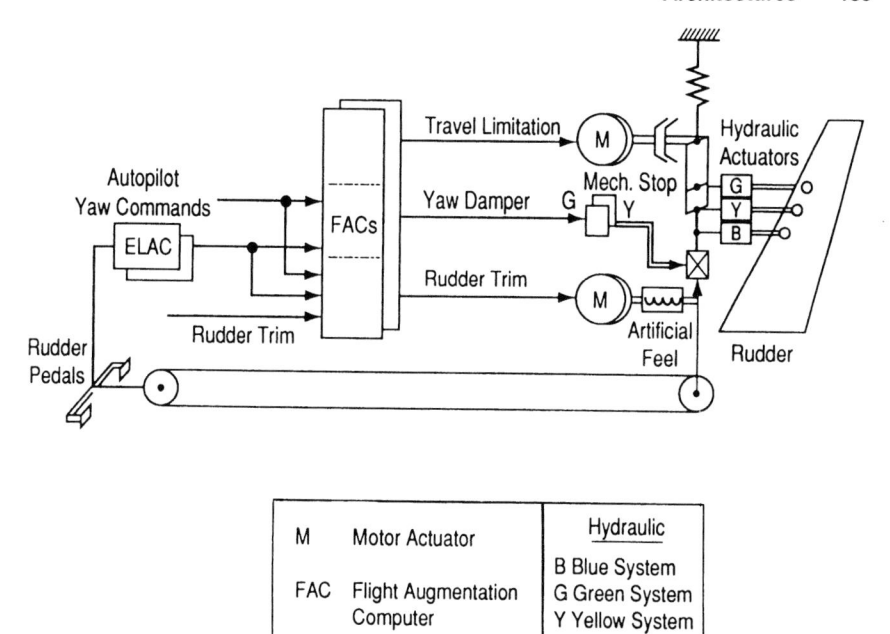

Figure 7.10c A-320 Yaw control system. (*Courtesy of Airbus Industrie.*)

are listed in Table 7.2 under the broad categories of sensors, data acquisition and distribution, and data processors, indicators, and actuators.

Figure 7.11a is a top-level view of the architecture showing how the subsystem elements were assembled into an integrated system and the principle interfaces (via ARINC 429 data buses) between the integrated avionics processor system (IAPS), sensors, data acquisition unit (DAU), and the many displays. Note the provisions for crew input to the IAPS via the control wheels and the control display units (CDUs).[7]

The IAPS, shown in Fig. 7.11b, is the heart of the system. It contains dual flight control computers (FCCs) for autopilot functions and dual flight management computers (FMCs) for flight guidance and management functions. The IAPS also contains a pair of dual input-ouput concentrators (IOCs). In order to achieve the needed data redundancy and integrity, each pair of IOCs is connected to a different set of sensors. The IOCs receive the data from the large number of sensors on the aircraft and then transform it into formats that are optimized for the selected user subsystem, such as the EFIS or the FMC.[7] All of the IAPS software is written in Ada. The electronic flight display (EFD) has 31,000 lines of source code occupying 318,000 bytes of target memory, and the FMS has 51,000 lines of source code requiring

TABLE 7.2 Subsystem Elements of the Starship Avionics System[7]

Sensors	
Navigation sensors 　VHF navigation receivers 　Distance measuring equipment 　VLF/Omega receivers w/H-field 　　antenna 　ADF receivers w/single or dual antenna 　Radio altimeter receiver/transmitter 　　w/antennas 　Global positioning system receivers— 　　provisions 　Microwave landing system receivers— 　　provisions Communications subsystem 　VHF communications transceivers 　Radio tuning units (RTUs) 　ATC transponders 　Audio subsystems	Weather radar sensor 　Weather radar transmitter/receiver/ 　　antenna—integrated assembly Air data subsystem 　Air data computers 　Altitude/vertical speed/preselector 　　indicators (ALIs) 　Airspeed/SAT/TAS/TAT indicators 　　(ASIs) 　Pneumatic altimeter and airspeed 　　indicators Attitude/heading sensors 　Strapdown attitude/heading reference 　　sensors 　Flux detectors 　Standby gyro horizon

Data Acquisition and distribution	
Aircraft data acquisition subsystem (ADAS) 　Data acquisition unit (DAU) 　Engine data concentrators (EDC)	Integrated Avionics Processor Subsystem (IAPS) 　Input-output concentrators (IOCs) 　Power supply modules 　IAPS card cage assembly

Data processors, indicators, and actuators	
Flight control system (FCS) 　Flight guidance/autopilot modules 　　(FCC)—p/o IAPS 　Primary servos 　Trim servo 　Autopilot control panel 　Flight guidance mode control panels Flight management system (FMS) 　Navigation and guidance processor 　　modules—p/o IAPS 　Control/display processors and database 　　modules—p/o IAPS 　Database unit (installation in aircraft 　　is optional) 　Control display units (CDUs) 　Altitude awareness panels 　Course/heading and multifunction 　　control panel	Integrated display system 　Primary flight displays (PFDs) 　Navigation displays (NDs) 　Engine indication, caution, and advi- 　　sory subsystem (EICAS) 　Multifunction display (MFD) 　Sensor display driver 　Sensor display units (SDUs)

Architectures 137

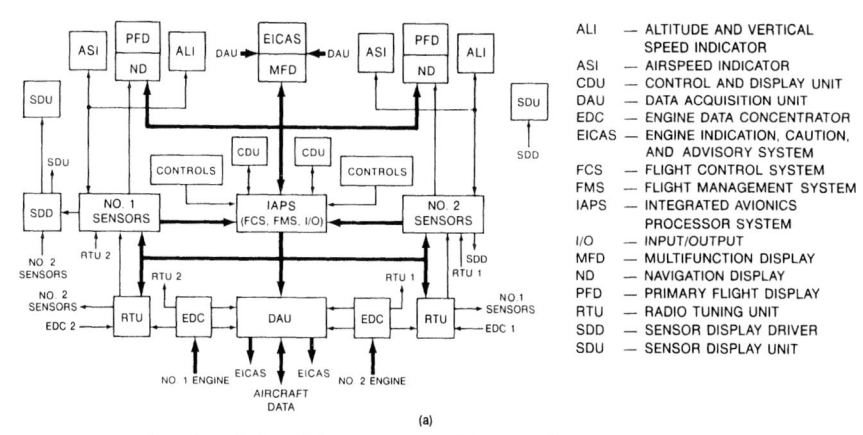

Figure 7.11a Beechcraft Starship avionics architecture.[7]

ALI	— ALTITUDE AND VERTICAL SPEED INDICATOR
ASI	— AIRSPEED INDICATOR
CDU	— CONTROL AND DISPLAY UNIT
DAU	— DATA ACQUISITION UNIT
EDC	— ENGINE DATA CONCENTRATOR
EICAS	— ENGINE INDICATION, CAUTION, AND ADVISORY SYSTEM
FCS	— FLIGHT CONTROL SYSTEM
FMS	— FLIGHT MANAGEMENT SYSTEM
IAPS	— INTEGRATED AVIONICS PROCESSOR SYSTEM
I/O	— INPUT/OUTPUT
MFD	— MULTIFUNCTION DISPLAY
ND	— NAVIGATION DISPLAY
PFD	— PRIMARY FLIGHT DISPLAY
RTU	— RADIO TUNING UNIT
SDD	— SENSOR DISPLAY DRIVER
SDU	— SENSOR DISPLAY UNIT

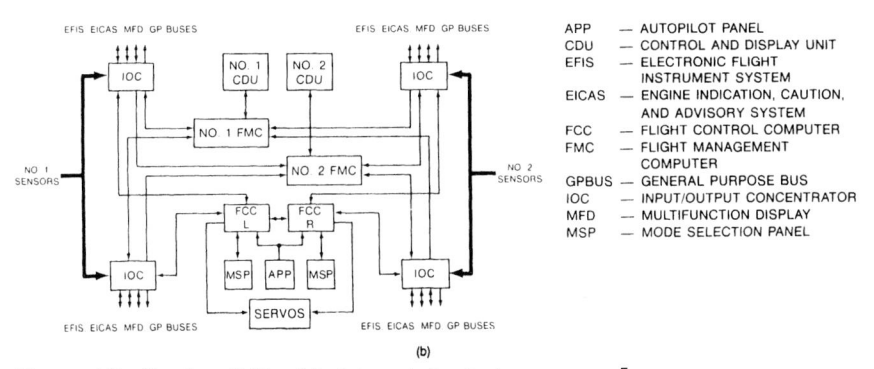

Figure 7.11b Beechcraft Starship integrated avionics processor.[7]

APP	— AUTOPILOT PANEL
CDU	— CONTROL AND DISPLAY UNIT
EFIS	— ELECTRONIC FLIGHT INSTRUMENT SYSTEM
EICAS	— ENGINE INDICATION, CAUTION, AND ADVISORY SYSTEM
FCC	— FLIGHT CONTROL COMPUTER
FMC	— FLIGHT MANAGEMENT COMPUTER
GPBUS	— GENERAL PURPOSE BUS
IOC	— INPUT/OUTPUT CONCENTRATOR
MFD	— MULTIFUNCTION DISPLAY
MSP	— MODE SELECTION PANEL

470,000 bytes of target memory. The software runs on Intel 80186 processors.[8]

One of the most notable features of the IAPS is its packaging. The IAPS cabinet is divided into identical top and bottom halves. It is mounted in the bottom of the coat closet behind the co-pilot and measures 11.4 in wide, 19.7 in high, and 6.0 in deep. The cabinet is populated with modules for each of the major functions: power conditioning (dual), flight management (dual), flight control, and IOC (dual). Because of the careful design, the use of state-of-the-art components, and modular packaging, a selected portion of the Starship equipment weighs 96.5 lb and occupies 2.9 ft^3, while the equivalent avionics using ARINC 600 packaging and ARINC 700 series equipment characteristics weighs 243 lb and occupies 6.5 ft^3.[8]

References

1. Dietrich, A. J., and Thomas, F. J., "Digital Computer Architecture as Applied to an Advanced Flight Control System," AIAA paper no. 85-1949-CP, 1985.
2. AFWAL-TR-87-1114 "Architecture Specification for Pave Pillar Avionics," January 1987.
3. Morgan, Michael J., "Integrated Modular Avionics for Next-Generation Commercial Airplanes," *IEEE Aerospace and Electronic Systems Magazine*, vol. 6, no. 8. pp. 9–12, August 1991.
4. Kelley, Michael R., "Airborne Computer Technology Initiatives," *Proceedings of the RTCA Annual Assembly*, December 3, 1990.
5. Davies, C. R., "Systems Aspects of Applying Active Control Technology to a Civil Aircraft," *Active Control Technology Experience and Prospects*, Royal Aeronautical Society, May 13–14, 1987.
6. Corps, S. G., "Airbus A320 Side Stick and Fly-by-Wire—An Update," *SAE Aerotech '86*, SAE paper 861801, 1986.
7. Fenwick, Charles A., and Spencer, James L., "Avionics System for the Beechcraft Starship 1," *SAE Aerotech '85*, SAE paper 851851, 1985.
8. Funk, David W., "Applying Ada to Beech Starship Avionics," *NASA Johnson Space Center First International Conference on Ada Programming Applications for the Space Station*, 1986.

Bibliography

Traverse, P., "Airbus and ATR System Architecture and Specification," in *Software Diversity in Computerized Control Systems*. Springer-Verlag Wien, New York, 1988.

Chapter

8

Packaging and Fitting the System into the Aircraft

One of the toughest parts of building a good digital avionics system is fitting the system into the aircraft. This is where the electrons meet the sheet metal and where a whole new breed of issues are encountered.

The most common image of avionics is the familiar "black box," more formally known as an LRU. Every digital avionics system usually ends up packaged in one or more LRUs. This chapter presents descriptions of standardized LRUs for both civil and military avionics. These standard LRU designs offer many advantages to both the designer and the operator in the form of interchangeability, easier maintenance, and a clement operating environment for the avionics.

In the opinion of many experts, keeping the avionics cool is the secret to highly reliable operation.* This chapter looks at some avionics cooling methods and LRU design techniques that reduce thermal stress on the components.

Perhaps the toughest part of installing an avionics system in an aircraft is the interfaces. The final section of this chapter interprets interfaces in two parts: what the aircraft provides to the avionics and what the avionics impacts on the aircraft are.

Civil Packaging Standards

ARINC Specification 600 is the premier example of an avionics standard that really works. ARINC 600 establishes the form factor, exter-

---

*Other experts believe that thermal cycling, not the maximum temperature, is the cause of poor reliability (see Chap. 6).

nal design, and environmental interfaces with the host aircraft for LRUs installed on civil aircraft.[1] It also provides a common language for describing the size and shape of any LRU and ensures that an LRU manufactured by one vendor to perform a given function can be exchanged with one performing the same function manufactured by another vendor. This interchangeability has enormous operational and economic benefits for the aircraft operators. Therefore, they fervently adhere to the design philosophy in the specifications and expect all avionics vendors to do the same.

It is important to note that ARINC 600 does *not* specify the internal design or performance details of the LRU, only the form factor of the box and the environment in which it will operate.* By focusing only on the top-level design description, the aircraft operators, who write the specifications, are spurring competition among the avionics manufacturers, which leads to a superior product.

Figure 8.1 shows the basic design for an ARINC 600 LRU.† All LRUs are 194 mm (7.64 in) high and 324 mm (12.76 in) deep, which is the same as an ARINC 404‡ ATR§ short case.

ARINC 600 introduces modular concept unit (MCU) as the basic unit of width for an LRU. Eight MCUs are equal in width to one ATR as defined in ARINC 404, or approximately 25.4 cm (10 in). The actual width of an LRU can be determined by the formula

$$W = (N \times 33) - 8.0 \text{ mm}$$

or

$$W = (N \times 1.30) - 0.32 \text{ in}$$

where W = width and N = number of MCUs. The rack space required in the avionics bay by an LRU is simply $N \times 33$ mm (or $N \times 1.30$ in).

Electrical connections are located on the back panel of the LRU. Electrical power must meet the requirements of RTCA document DO-

---

*LRU performance specifications, such as resolution, accuracy, update rate, reliability, and so on, are generally established in Minimum Operational Performance Standards (MOPS) or similar documents published by the RTCA or equivalent organizations.

†ARINC 600 is written in International System units. All dimensions will be given in those units first, followed by the English units.

‡ARINC 404 Air Transport Equipment Cases and Racking is used for designing analog equipment packaging and is mentioned here only to show the heritage of ARINC 600.

§ARINC 404 established Air Transport Racking (ATR), but on April 11, 1967, the Airlines Electronic Engineering Committee corrected ARINC 404 so that ATR came to mean Austin Trumbull radio as was originally intended to honor the founder of the concept of standardized packaging and interfaces for air transport avionics.

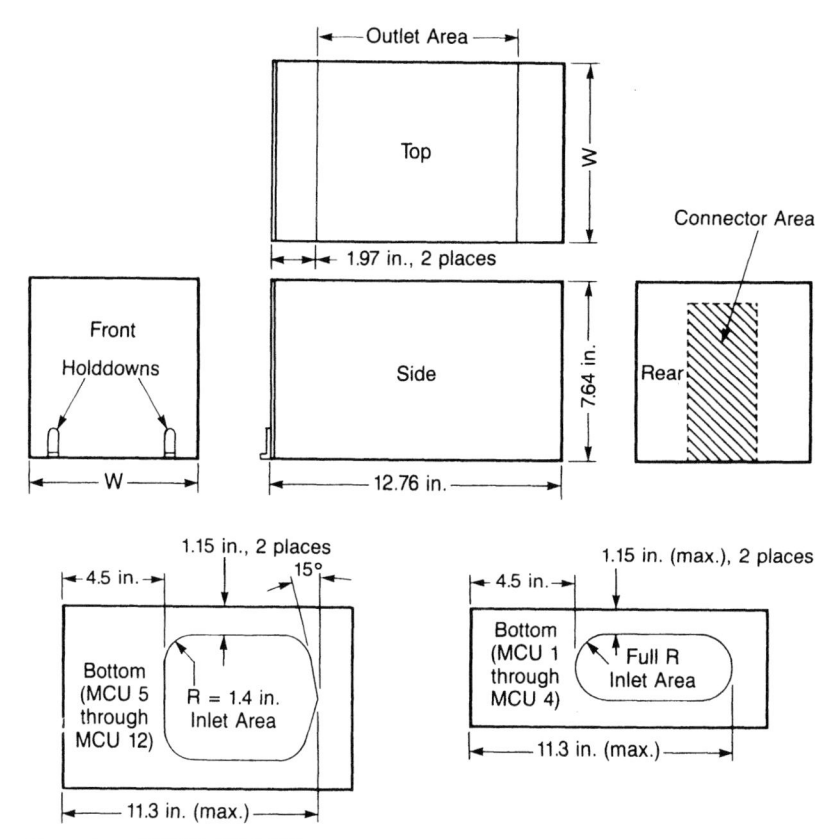

Figure 8.1 ARINC 600 LRU.

160 or its equivalent EUROCAE document ED-14, sections 16.3 and 17. (See Chap. 4 for a discussion of power quality requirements.)

Allowable power dissipation is a function of the LRU width. For LRUs *without* forced air cooling, power is limited to $5 + 2.5N$ W. Thus, for a one-MCU-wide LRU without forced air cooling, power dissipation is limited to 7.5 W and for a five-MCU-wide LRU to 17.5 W. For LRUs *with* forced air cooling, power dissipation is much higher and can reach 25 W/MCU. Regardless of the allowable power dissipation, the exterior vertical surfaces of the LRU must not exceed 60°C (140°F) average or 65°C (149°F) at a hotspot.

The temperature of cooling air for continuous operation can range from −15°C (5°F) to 55°C (131°F). As a design point, if 30°C (86°F) cooling air is used in flight, the flow rate must be 136 kg/(h)(kW) [300 lb/(h)(kW)]. The air must not contain particles larger than 400 micrometers, or μm (0.016 in).

Cooling air openings, if used, are on the top and bottom of the LRU. Air flow is upward.

An optional requirement for equipment with forced air cooling is normal operation of the equipment, without cooling air, for up to 90 min in ambient air that rises linearly in temperature from 45°C (103°F) to 60°C (140°F) during the period.*

The weight of an ARINC 600 LRU is limited to 2.5 kg (5.5 lb)/MCU, not to exceed 20 kg (44 lb). Thus, a four-MCU-wide LRU could weigh up to 10 kg (22 lb) and a ten-MCU-wide LRU, up to 20 kg (44 lb).

ARINC 600 has been adapted, with some changes, for use on large military aircraft as DOD-STD-1788 Avionics Interface Design Standard. This standard is discussed in the next section.

Military Packaging Standards

As mentioned in the previous section, DOD-STD-1788 Avionics Interface Design Standard establishes the form factor, mounting, and environmental operating conditions for avionics to be installed in the avionics bay of large military aircraft.[2] This standard is based on ARINC 600 and its predecessor, ARINC 404, and they share many similarities. Military avionics designers are required to use 1788 wherever possible to achieve the reliability, maintainability, and life cycle cost benefits promised from its application.

Like its civil predecessor, DOD-STD-1788 Avionics Interface Design Standard establishes standard LRU case dimensions. A drawing of a standard case is shown in Fig. 8.2. All cases have a standard height of 194 mm (7.64 in) and a depth of 324 mm (12.76).†

DOD-STD-1788 describes the width of an LRU by a size number, N. The actual width of an LRU is determined by the formula

$$W = (N \times 33) - 8.0 \text{ mm}$$

or

$$W = (N \times 1.30) - 0.32 \text{ in}$$

where W = width and N = LRU size number. The rack space required in the avionics bay by an LRU is simply $N \times 33$ mm (or $N \times 1.30$ in).

Long-standing terms used to describe the dimensions of civil avionics LRUs are MCU, as established by ARINC 600 Air Transport Avionics Equipment Interfaces, and Austin Trumbull radio (ATR), as es-

---

*One manufacturer of extended range operation passenger transport aircraft requires that the avionics operate without cooling air for up to 18 h.

†DOD-STD-1788 is written in International System units. All dimensions will be given in those units first, followed by the English units.

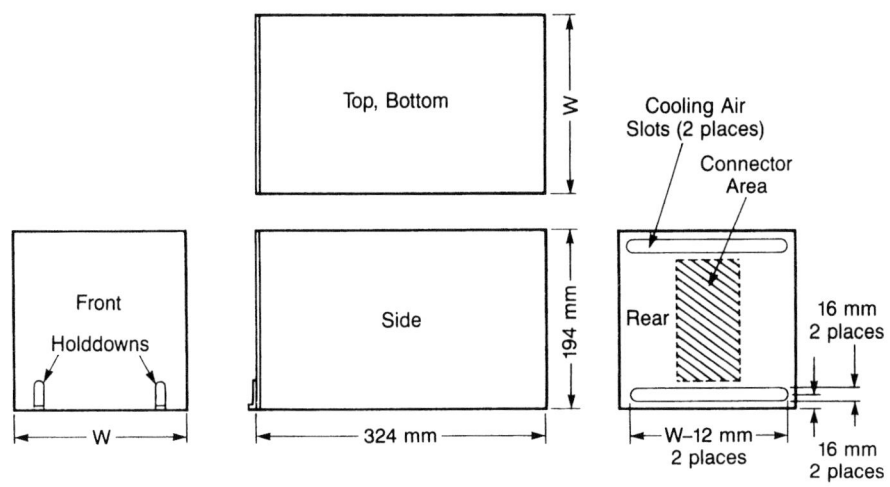

Figure 8.2 DOD-STD-1788 LRU.

tablished in ARINC 404 Air Transport Equipment Cases and Racking. The size number, N, of a 1788 LRU matches the width of a civil LRU in MCUs. Thus, a 1788 LRU where $N = 8$ is identical in width to an ARINC 600 LRU where the width is eight MCUs. The depth of a 1788 LRU is the same as the depth of an ARINC 600 LRU.

Similar comparisons can be made between 1788 and ARINC 404. A 1788 LRU where $N = 8$ is equal in width to a one-ATR-wide ARINC 404 LRU. The standard 1788 LRU depth of 324 mm (12.76 in) equals a short ATR case and an alternate depth of 502 mm (19.76 in) equals a long ATR case.

The rear panel has cooling air inlets at the top and bottom and also contains the electrical connector(s). An alternate case design, referred to as the low-profile tray, but never discussed further in the standard, limits the case height to 170 mm (6.7 in) and places the cooling air inlets on the side of the rear panel. For either case design, the cooling air exits the LRU through slots in the front panel. However, if a closed cooling system is used, there are no front openings; the rear panel top slot is the air inlet and the bottom slot is the return.

DOD-STD-1788 goes beyond specifying the dimensions of an LRU case to present many other important design considerations. They include constraints on the weight and cooling air consumption as a function of the width of the LRU. The weight is limited to 3.5 kg (7.7 lb)/N, not to exceed a maximum of 27.5 kg (60.5 lb). In other words, LRUs where $N \geq 8$ cannot weigh more than 27.5 kg (60.5 lb).

Power dissipation for LRUs *with* cooling air is limited to 125 W/N. Thus, an LRU for which $N = 2$ can dissipate up to 250 W and an LRU for which $N = 12$ can dissipate up to 1500 W. LRU power dissipation *without* cooling air is limited to $10 + 2.5 \times (N - 2)$ W. For uncooled LRUs, then, the maximum power dissipated for an LRU for which $N = 2$ is 10 W and where $N = 12$ is 35 W.

In the case of loss of cooling air to an LRU for 5 min or less, the maximum temperature of microcircuits must not exceed 125°C (257°F) and power devices must not exceed 150°C (302°F), regardless of the power dissipation permitted in the LRU.

Cooling air at the inlet can range in temperature from −57°C (−71°F) to 47°C (117°F). At 30°C (86°F) the flow rate should be 58 kg/(h)(kW) [128 lb/(h)(kW)].

In designing the LRU, cooling air is not permitted to flow directly over electronic component to keep it at normal operating temperatures. The maximum temperature of an LRU side panel is limited to 76°C (169°F) to minimize radiative or convective heat transfer to adjacent LRUs.

DOD-STD-1788 requires the LRU design to be "thermally optimized within appropriate design constraints to minimize the LRU life cycle cost (LCC) and optimize the LRU reliability (based on the predicted reliabilities of the individual parts)." The results of the analysis should be used to optimize the location of piece parts to improve their reliability. Guidance on locating parts within an LRU to optimize thermal performance is discussed further in the next section.

A comparison of selected LRU parameters in ARINC 600 and DOD-STD-1788 is given in Table 8.1. Note the relatively short operating time and substantially higher maximum external case temperatures, without cooling air, for 1788 when compared to ARINC 600. Thus, a source of cooling air that can be brought on line quickly is required for military avionics.

The Pave Pillar and B-777 AIMS integrated modular avionics architectures (see Chap. 7) require a different packaging concept from the familiar LRUs to achieve their promises of lighter weight and reduced maintenance. In response to this requirement line replaceable modules (LRMs) have been developed that are essentially ruggedized circuit boards with added provisions for mounting and cooling. These modules are mounted in a cabinet in the avionics bay that provides a relatively benign operating environment and electromagnetic shielding.

The Pave Pillar LRMs are based on MIL-STD-28787 Modules, Standard Electronic, General Specification for.[3] MIL-M-28787 describes a number of standard configurations and sizes for electronic modules. Standard Electronic Module Size E (SEM-E) form factor has been chosen for the Pave Pillar. A SEM-E module is 5.88 in high and 6.4 in deep. The width of the module can vary in 0.1 in increments from a minimum of 0.28 in to a maximum of 0.58 in. Most of the Pave Pillar modules are 0.38 in wide.

TABLE 8.1 Comparison of Selected LRU Characteristics in ARINC 600 and DOD-STD-1788

	ARINC 600	DOD-STD-1788
Case dimensions		
Height	194 mm (7.64 in)	
Depth	324 mm (12.76 in)	
Width	$N \times 33 - 8$ mm ($N \times 1.3 - 0.32$ in)*	
Cooling air flow rate, 30°C	136 kg per hr per kW (300 lbs per hr per kW)	58 kg per hr per kW (128 lbs per hr per kW)
Cooling air flow, °C	Bottom to Top	Rear to Front
30	136/5.0†	88/3.2†
40	220/8.1†	117/4.2†
Emergency operation without cooling air	90 min	5 min
Maximum external case temperature	60°C (140°F) [60°C (149°F) locally]	76°C (169°F)
Maximum weight	20 kg (44 lbs)	27 kg (60 lbs)
Maximum power dissipation		
with cooling air	25 W per N	125 W per N
without cooling air	$5 + 2.5 \times N$ W	$10 + 2.5 \times (N - 2)$ W

*For ARINC 600 N = number of MCUs, for DOD-STD-1788 N = size number.
†Values are shown $xxx/y \cdot y$ where xxx is in kg/h $\cdot$ kW and $y \cdot y$ is in lb/min $\cdot$ kW.

A number of Pave Pillar SEM-E LRMs are available including a MIL-STD-1750A processor, volatile and nonvolatile bulk memories, a MIL-STD-1553B bus interface processor, and a dc-dc converter. There are many other types of modules also available for other applications. For each type of SEM-E LRM, a supplement to MIL-M-28787 is issued describing the performance and physical characteristics of the LRM including connector pin assignments and key pin configuration.

Keeping the Avionics Cool

Conventional wisdom says that all electronics should be kept cool to achieve long life. This is especially true for avionics, which are frequently inconvenient and expensive to maintain.

In the previous two sections of this chapter standards for interfacing the avionics with the aircraft, including the need for cooling air, were discussed. This section focuses on the benefits of keeping the avionics cool and offers design techniques to reduce thermal stress on the components.

Many studies have demonstrated the value of reducing the temperature of electronic components to extend their lifetime. Typical results from these studies are shown in Fig. 8.3. This figure shows that a 10°C

146 Chapter Eight

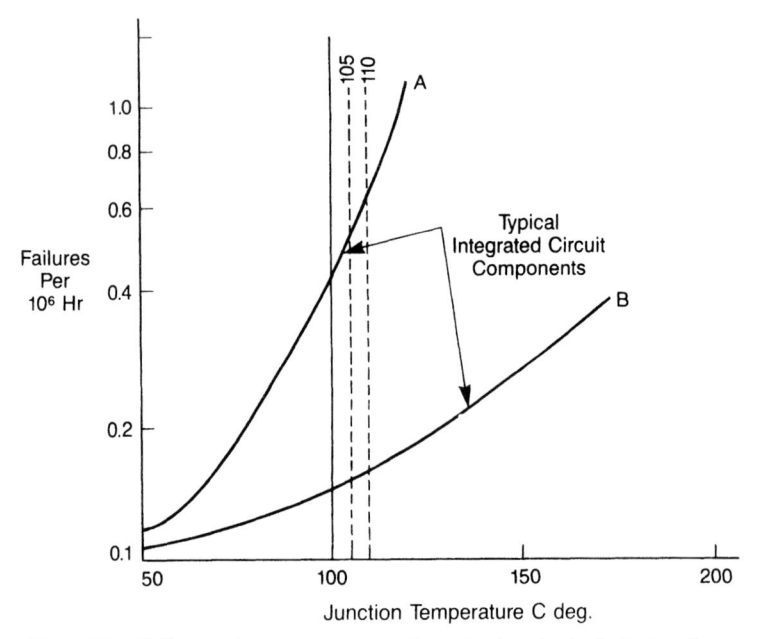

Figure 8.3 Failure rate versus temperature for two typical electronic components. (*Reprinted with permission.* © *1983 Society of Automotive Engineers, Inc. SAE Paper No. 831107 by Robert L. Berger.*)

(18°F) rise in junction temperature increases the device failure rate by about 50 percent for a component that matches curve A. Raising the temperature of a component that matches curve B by 10°C (18°F) raises the failure rate by only about 10 percent.

Passive cooling is a simple and obvious way to keep the avionics cool. It has been used successfully in the past, but all modern avionics, especially in densely packed military LRUs, have required active cooling, that is, air being pulled over the avionics by fans. Reexamination of passive cooling by the Boeing Commercial Airplane Co. has shown that properly designed modules in a modular avionics architecture can be adequately cooled by passive means. Passive cooling saves weight and improves the overall reliability of the aircraft since there is no avionics cooling equipment to fail. Figure 8.4 shows the difference in case temperature and air temperature for a typical component on a circuit board as a function of power being dissipated on the board. The solid circle depicts the current ARINC 600 cooling performance. The open square and solid square curves depict two cases of passive cooling; the difference between the two curves is that the solid square curve shows the effect of adding cooling fins ("thermal ladders") to the highest power dissipating components.

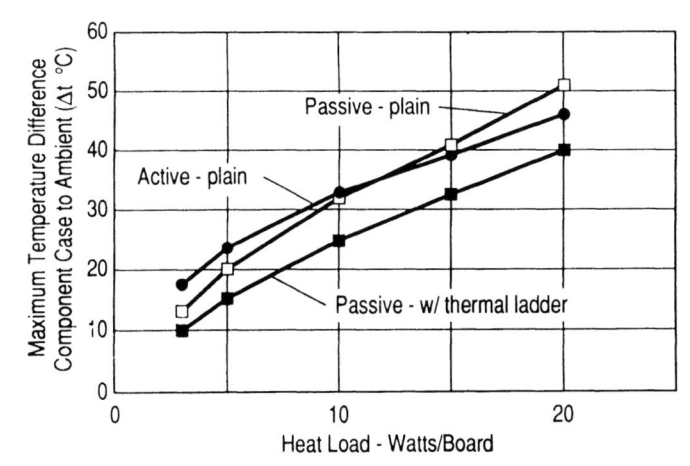

Figure 8.4 Comparison of maximum temperature rise over ambient for actively (ARINC 600) and passively cooled components. (*Courtesy of the Boeing Commercial Airplane Co., Charles T. Leonard. Used with permission.*)

ARINC 600 and DOD-STD-1788 recommend the use of cooling air to keep the entire LRU at or below a selected operating temperature limit. The more competent and astute designers go beyond these recommendations and conduct analyses and tests to optimize the location of components on circuit boards to substantially reduce component and, therefore, LRU failure rates. The U.S. Air Force has established a Thermal Management Control (TMC) program as Task 106 of MIL-STD-785 Reliability Program for Systems and Equipment Development and Production.[4] The objective of the TMC program is to allocate cooling capacity at the system level to minimize LCC. A subsidiary program, Electronic Equipment Thermal Management, requires the avionics equipment design to be optimized for minimum avionics LCC, not just to obtain a simple reliability goal. If an LRU contained a circuit board with two components, one of which had a failure rate versus temperature that matched curve A in Fig. 8.3 and the other component matched curve B, emphasis should be placed on reducing the operating temperature of the curve A type of component since a reduction in temperature for it will yield the largest reduction in the LRU failure rate. The most common ways to reduce the temperature of a component are to relocate it on the circuit board or to relocate the circuit board within the LRU relative to the cooling air.

Heat pipes are another promising method of cooling avionics. Heat pipes are closed containers with a working fluid and a capillary wick. As heat is absorbed in the hot end of the pipe, it vaporizes the working fluid, which is driven to the cold end where the heat is removed and the vapor returns to the liquid state. The liquid is then transported by capillary action back to the hot end of the pipe where the process is repeated. Tests were conducted with typical avionics circuit boards containing bulk memory devices—one board with an aluminum substrate and the other with a heat pipe substrate. The results showed a 16°C (29°F) case temperature variation among components at different locations on the aluminum substrate board but only a 3°C (5.4°F) case temperature variation among those components on the heat pipe substrate board. Those components on the heat pipe substrate board were also substantially cooler, which translated into an estimated 40 to 60 percent greater mean time between failure (MTBF).[5]

One major airline cools the entire avionics bay with a separate onboard air conditioner any time the avionics are operating on the ground and the bay temperature is higher than 26°C (80°F). The MTBF of the LRUs is improved by approximately 60 percent. The reduced maintenance cost resulting from the improved MTBF more than offsets the weight, complexity, and cost penalties of the additional air conditioning.

Specifying the Interfaces with the Aircraft

Another major task early in the design of avionics systems is the definition of interfaces with the host aircraft. Interfaces are a two-part situation where what the aircraft provides to the avionics is as important as what the avionics impacts are on the aircraft.

Interface specifications describe the aircraft environment in precise, quantitative terms for the avionics designer. Specifications are usually provided for power, cooling and ambient air, ambient pressure, mounting plate temperatures, electromagnetic environment, shock, vibration, volume, and instrument panel size and shape. In many cases, the aircraft requirements are stated in military standards or specifications, RTCA documents, or ARINC characteristics, in lieu of a detailed description of the interface. Typical of these standards are MIL-STD-704 Aircraft Electrical Power Characteristics and RTCA document DO-160 Environmental Considerations and Test Procedures for Airborne Equipment for aircraft electrical power (see Chap. 4). Many of the environmental specifications are covered by MIL-STD-810 Environmental Test Methods and Engineering Guidelines or by RTCA document DO-160 mentioned earlier (see Chap. 9).

Where an existing standard is used in lieu of a detailed interface specification, the abbreviated specification will frequently tailor and/or alter the existing standard to meet the unique needs of the host aircraft. In the interests of cost and schedule, alterations to standard procedures and conditions should be held to a minimum.

Retrofitting avionics to older, in-service aircraft requires that special attention be given to the interfaces. The avionics system designer must have complete, detailed information about the host aircraft such as the amount of equipment space, power and cooling available, the location of bulkheads, cable routings, maintenance access points, and the space available for mounting new sensors. The designer must also consider the impact of removing the older equipment, or if the equipment is being retained, the impact on its performance and reliability when disturbed by the installation of the new avionics.

The specifications of the avionics system interfaces with the aircraft must contain information such as volume, weight, power requirements, electromagnetic emissions, wire routing, sensor locations, and alignment requirements. This specification is prepared by the avionics system designer for use by other system designers.

All interface specifications or description documents must be subject to rigid configuration control to ensure the integrity of the interface and that all changes to the interface are known to all interested parties.

References

1. ARINC 600 Air Transport Avionics Equipment Interface.
2. DOD-STD-1788 Avionics Interface Design Standard, 15 May 1986.
3. MIL-M-28787 Modules, Standard Electronic, General Specification for, 30 March 1989.
4. MIL-STD-785 Reliability Program for Systems and Equipment Development and Production, Notice 1, 3 July 1986.
5. Basiulis, A., and Minning, C. P., "Improved Reliability of Electronic Circuits through the Use of Heat Pipes," *Proceedings of the IEEE National Aerospace Electronics Convention,* 85 CH 2189-9, 1244-7, 1985.

Chapter

9

Hardware Assessment and Validation

Hardware assessment and validation are major parts of building a modern digital avionics system. It is essential that the designer make a critical examination of the proposed system at many stages of design and build. The earliest phases of assessment and validation focus on qualitative evaluation. Does the design make good engineering sense? Are the designer and other experts comfortable with it?

For civil aircraft early discussions with the regulatory agencies are necessary to avoid false starts and extensive work on a system that may have to be redesigned later to meet certification concerns.

Quantitative evaluation of the system relies on proven analysis techniques such as fault tree analysis (FTA) and failure modes and effects analysis (FMEA). These techniques are complementary and collectively provide an overall view of the performance of the system, especially under off-nominal conditions.

Hardware assessment also includes the familiar "shake and bake" environmental tests to evaluate performance under a number of environmental stresses, both individually and combined. The issue of electromagnetic interference must also be examined.

For complex, fault-tolerant systems quantitative analysis requires the use of modern computer-based tools. Because of the seemingly infinite number of states a fault-tolerant system can assume, a thorough analysis mandates the need for computer assistance.

The ilities such as reliability, maintainability, affordability, survivability, and certificability are also good factors to keep in mind when assessing a system. A good designer recognizes the importance of every ility on a seemingly endless list.

Tips on Qualitatively Evaluating Systems

One of the major goals of this text is to provide guidance for evaluating digital avionics systems architectures, and a substantial portion of the book is devoted to this topic. The present discussion is intended to only introduce the most common factors to be considered in an evaluation and establish their relative importance. The reader is directed to other sections of this chapter and to other chapters for detailed, quantitative techniques for evaluating systems.

There are almost as many evaluation factors for systems architectures as there are system architects. The selection of evaluation factors varies with the application of the system; however, every list should include, as a minimum, the following:

Capability

Reliability

Maintainability

Certificability (civil)

Life cycle cost (military) or cost of ownership (civil)

Survivability (military)

Technical risk

Weight

Power

The exact priority order of the list is always an issue. However, traditionally, the three most important factors in all cases are capability, reliability, and maintainability. Beyond these three items, the order and length of the list will vary widely depending on the system and its application. The most important ility is capability. How capable are the avionics? Can they do the job and even more? It is the task of the designer to maximize the capability of the system within the constraints that are imposed.

Reliability is one of the most important factors in evaluating an architecture. Every designer strives to make systems as reliable as possible since higher reliability generally leads to lower maintenance costs. (There are times, however, when this may not hold true. It is possible that the designer may spend so much time developing a highly reliable design with exceptionally reliable parts that the system development and acquisition costs can be excessive.) Reliability begins with the piece part reliability rates found in documents such as MIL-HDBK-217: Reliability Prediction of Electronic Equipment. Beyond that, the computation and evaluation of system reliability becomes very complex as the nuances of fault-tolerant designs are considered. Computer-based models are required to determine the reliability of highly integrated, fault-tolerant systems. But the bottom line remains the same as for earlier, simpler designs: No matter how elegant the system architecture is, if it is not reliable, the customer will not buy it, and, in the case of civil avionics, the regulatory agencies may not certify it.

Closely related to reliability in importance is maintainability. Every system will eventually need preventive or corrective maintenance. Thus, careful attention must be paid to how easily the system can be maintained through built-in testing, automated troubleshooting, and easy access to the hardware. The designer may want to consider the trade-off between accessibility and reliability: If the equipment can be proven to be extremely reliable, it can be placed in remote, relatively inaccessible locations that, despite this inaccessibility, enhance the overall system performance.

The combination of reliability and maintainability is availability. Trade-offs can be made between reliability and maintainability to optimize availability. Availability translates into sorties for military aircraft and into revenue flights for civil applications.

Certificability is a major area of concern for avionics in civil aircraft. Certification, conducted by the regulatory agencies, is based on detailed, expert examination of *all* facets of the aircraft design and operation. In order to expedite certification, the avionics architecture should be straightforward and easily understandable. There should be no sneak circuits and no nonobvious modes of operation. The early stages of avionics certification focus on three analyses: preliminary hazard, fault tree, and failure modes and effects. These analyses must be conducted as early as possible in the design process and approved by the regulatory agency.*

Life cycle cost (LCC) for military applications and cost of ownership (COO) for civil applications are the principle economic measures for evaluating avionics architectures. Included in these measures are costs for such varied items as spares acquisition, transportation and storage, training (both air crew and maintenance personnel), hardware development and test, depreciation, and interest. Achieving a credible economic evaluation of the architecture requires close cooper-

---

*In the United States the regulatory agency for civil aviation is the FAA. In the United Kingdom it is the Civil Aviation Authority (CAA), and in France it is the Direction Generale de l'Aviation Civile (DGAC). In other nations, the agency is referred to by other names. Throughout this book FAA should be interpreted as a generic term referring to the appropriate state-sponsored regulation and certification agency.

ation between the designer and the cost specialist. Each should apply his or her own expertise to ensure that no details are overlooked or misunderstood.

For military avionics the issue of survivability is another major concern. Survivability is defined as a function of susceptibility and vulnerability. Susceptibility is a measure of the probability that an aircraft will be hit by a given threat, and vulnerability is a measure of the probability that damage will occur if there is a hit by the threat.

Another subjective factor in evaluating an architecture is technology risk. There is a natural tendency by every designer to demonstrate technical prowess by including the latest technology in the system. While the latest technology usually offers enhanced performance, the downside is increased risk that the technology may be difficult to manufacture, may not work in the application at hand, and, over time, problems and previously unknown deficiencies may appear. Thus, every architecture should be evaluated for both developmental and long-term technological risks.

Finally, in evaluating an architecture, estimates of weight and power must be made. The need to minimize weight and power requirements are two of the fundamental concepts in avionics design. There are trade-offs to be made since the lightest or least power-consuming design may not be the optimum design when other factors are considered. Implicit in making these estimates is the assumption that the line replaceable units (LRUs) have already been configured and located in the aircraft so that data and power bus lengths and configurations are known.

Major Regulations for Civil Avionics Certification

Within the overall task of hardware assessment and validation, certification is perhaps the most difficult part for civil avionics designers. Certification is the challenging process of negotiation and compromise between the designers and the regulatory authorities buttressed by technical analysis and expertise on both sides. Federal Aviation Regulations (FARs) establish the requirements and spell out the procedures to be followed to demonstrate that the avionics are airworthy and safe for use on civil transports. For digital avionics systems, the most important regulation is FAR Part 25.1309: Equipment, Systems and Installation. There are others that affect digital avionics including Part 25.581: Lightning Protection; Part 25.671: Control Systems, General; and Part 25.672: Stability Augmentation and Automatic and Power-Operated Systems. In this section the focus will be on Part 25.1309.

Since FAR 25.1309 is so important to civil transport avionics systems, the first four paragraphs are stated below:[1]

(a) The equipment, systems, and installations whose function is required by this subchapter must be designed to ensure that they perform their intended functions under any foreseeable operating condition.
(b) The airplane systems and associated components, considered separately and in relation to other systems, must be designed so that
  (1) The occurrence of any single failure condition which would prevent the continued safe flight and landing of the airplane is extremely improbable, and
  (2) The occurrence of any other failure condition which would reduce the capability of the airplane or the ability of the crew to cope with adverse operating conditions is improbable.
(c) Warning information must be provided to alert the crew to unsafe system operating conditions and to enable them to take appropriate corrective action. Systems, controls, and associated monitoring and warning means must be designed to minimize crew errors which would create additional hazards.
(d) Compliance with the requirements of paragraph (b) of this section must be shown by analysis, and where necessary, by appropriate ground, flight, or simulator tests. The analysis must consider
  (1) Possible modes of failure, including malfunctions and damage from external sources
  (2) The probability of multiple failures and undetected failures
  (3) The resulting effects on the airplane and occupants, considering the stage of flight and operating conditions, and
  (4) The crew warning cues, corrective action required, and the capability of detecting faults.

A cursory examination of this regulation quickly leads to questions concerning the precise, quantitative meaning of the terms *extremely improbable* and *improbable*. To provide guidance on this and other questions about 1309, the Federal Aviation Administration (FAA) issued Advisory Circular 25.1309-1A.[2] While the regulation establishes requirements, the advisory circular interprets the requirements and presents techniques to demonstrate compliance with them. Like all FAA advisory circulars, 25.1309-1A is *advisory* only and presents one method for meeting the FAR. Other methods can be used by the avionics manufacturers, but they must be approved by the FAA prior to use as being functionally equivalent to the one in 25.1309-1A. 25.1309-1A addresses hardware only; software is addressed in DO-178A.

Table 9.1 summarizes the 25.1309-1A adjective, qualitative, and quantitative interpretation of "extremely improbable" and "improbable." The initial step in the certification process[2]

TABLE 9.1 Summary Interpretation of FAR 25.1309*

FAR 25.1309 term	Failure condition	Effect on performance	AC 25.1309-1A interpretation Qualitative probability	AC 25.1309-1A interpretation Quantitative probability	Equivalent DO-178A term
Extremely improbable	Catastrophic	Prevent continued safe flight and landing	Not anticipated to occur in the entire fleet operational life	$\leq 10^{-9}$	Critical
Improbable	Major	1. Significant reduction in safety margins or functional capabilities, some occupant discomfort; 2. Large reduction in safety margins or functional capabilities, adverse effects on occupants	Occasionally in the life of all aircraft of one type	$>10^{-9}$ but $<10^{-5}$	Essential
Probable†	Minor	No significant reduction in aircraft safety	Once per aircraft lifetime	$\geq 10^{-5}$	Nonessential

*Advisory Circular 25.1309-1A System Design and Analysis; June 21, 1988.
†"Probable" is not used in FAR 25.1309, but it is used in the Advisory Circular.

Is to conduct a functional hazard assessment (FHA) to identify and classify all potentially hazardous failure conditions and to describe them in functional and operational terms. An FHA is qualitative and is conducted using experienced engineering and operational judgment.... the classification of failure conditions should always be accomplished with consideration of all relevant factors; e.g., system, crew, performance, operational, external, etc.

The result of the FHA is a classification of all potential failures as "catastrophic," "major," or "minor" based on descriptions like those in Table 9.1. While these classifications are not formally approved by the FAA, they should be reviewed and discussed with them to ensure that they are correct in the eyes of the FAA since they are the basis for all subsequent certification activity.

For systems or functions that could generate a minor failure condition, an FHA should be sufficient, provided the subject system or function is physically and functionally isolated from other, more critical systems.

Systems or functions that could generate a major failure condition require substantially more analysis to prove their performance and reliability. Engineering experience and operational judgment may be sufficient in some cases where, though the failure could be major, the system or function under study is not complex and is based on existing similar equipment operating under similar conditions. For more complex and/or new design, systems or functions FMEA and/or FTA may be required.

Systems or functions that could generate a catastrophic failure condition require a more thorough analysis. The techniques and judgments applied for analyzing major failure conditions are repeated for catastrophic failure conditions but in a more rigorous and detailed manner. Additional emphasis is given to identifying common-cause failures and failures that could cascade from minor into catastrophic.

The important point to remember about certifying civil avionics is that it is a time-consuming process, built on equal parts of fact and judgment, with extensive interaction with the certifying agency.

Fault Tree Analysis Is a Proven and Accepted Technique

FTA is a well proven and widely accepted technique developed originally for the nuclear power industry for safety analyses of power plants. This rigorous, logically based technique has been documented in Nuclear Regulatory Commission report NUREG 0492 Fault Tree Analysis.[3] NUREG 0492 is of importance to avionics designers since the techniques described therein have been recognized by the FAA as acceptable for certifying avionics.[4] Much of the material in this section has been extracted from Refs. 3 and 4.

While FMEA and FHA are both inductive reasoning techniques that lead from the individual case to the general case, FTA is a functionally oriented, deductive approach, leading from the general to the specific case. That is, given an event such as a system malfunction, what are the potential causes? FTA is defined in NUREG 0492 as

> An analytical technique whereby an undesired state of the system is specified (usually a state that is critical from a safety standpoint) and the system is then analyzed in the context of its environment and operation to find all credible ways in which the undesired event can occur. The fault tree itself is a graphical model of the various parallel and sequential combinations of faults that will result in the occurrence of the predefined undesired event.

Although initially intended only as a qualitative model, FTA analysis has become quantitative, especially in its application to digital avionics.

FTA, when applied to digital avionics, can serve several functions, including (1) to assure that the failure of no single component can cause the entire system to fail, (2) to identify critical modules in critical functions, and (3) to verify the adequacy of fault detection and recovery schemes. The technique, however, does have limits and becomes unwieldy when applied to a circuit at or below the board level.

An FTA is usually sufficient for analyzing critical and essential systems or functions if acceptable fault management can be demonstrated. An FMEA may be needed to quantify the probability of an undesired event.[5]

All fault trees, regardless of complexity, are constructed of a basic set of symbols, as shown in Fig. 9.1. The similarity to standard logic symbology is obvious. Simple, generic examples of the use of some of the symbols are given in Fig. 9.2.*

Figure 9.3 shows a top-level fault tree from the FAA *Digital Systems Validation Handbook*.[4] The top-level (Level 1) event is an unacceptable deviation from the path/attitude/speed profile during a crucial flight phase. The probability of that event occurring has been determined from the lower-level fault trees to be 0.638×10^{-9}. The Level 1 event could be caused by any one of three Level 2 events (note the OR logic symbol). It is important to note that the left Level 2 event

---

*In NUREG 0492 there are several terms that are used in a different way than in most other avionics documents. In NUREG 0492, a fault is a higher-order event that can be broken down into basic events known as failures. A failure is the lowest identifiable level of abnormal occurrence. Thus, the digital avionics designer who uses NUREG 0492 should be carefully attuned to this difference in terminology.

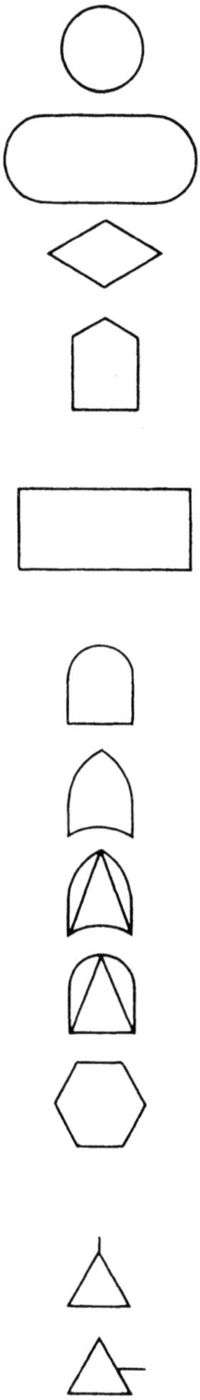

Primary Events

BASIC EVENT: The lowest possible level of definition, the limit of resolution.

CONDITIONING EVENT: Shows any conditions or restrictions that apply to the logic gate to which it is attached.

UNDEVELOPED EVENT: A fault event that is not developed further either because it is of insufficient importance or information is unavailable.

EXTERNAL EVENT: An event that is normally expected to occur but is not a fault.

Intermediate Events

INTERMEDIATE EVENT: A fault event that resulted from more antecedent causes acting through logic gates.

Gates

AND: An output event occurs only if all of the input events have occurred.

OR: An output event occurs if one or more of the input events occurs.

EXCLUSIVE OR: Output fault occurs if exactly one of the input faults occurs.

PRIORITY AND: Output fault occurs if all of the input faults occur in a specific sequence (as specified in a CONDITIONING EVENT shown at the right of the gate.)

INHIBIT: Output fault occurs if the single input fault occurs in the presence of an enabling condition (as specified in a CONDITIONING EVENT shown at the right of the gate.)

Transfer Symbols

TRANSFER IN: Indicates that the tree is continued in another location.

TRANSFER OUT: Indicates that this portion of the tree must be attached to the corresponding TRANSFER IN at another location.

Figure 9.1 Fault tree symbols.[3]

160   Chapter Nine

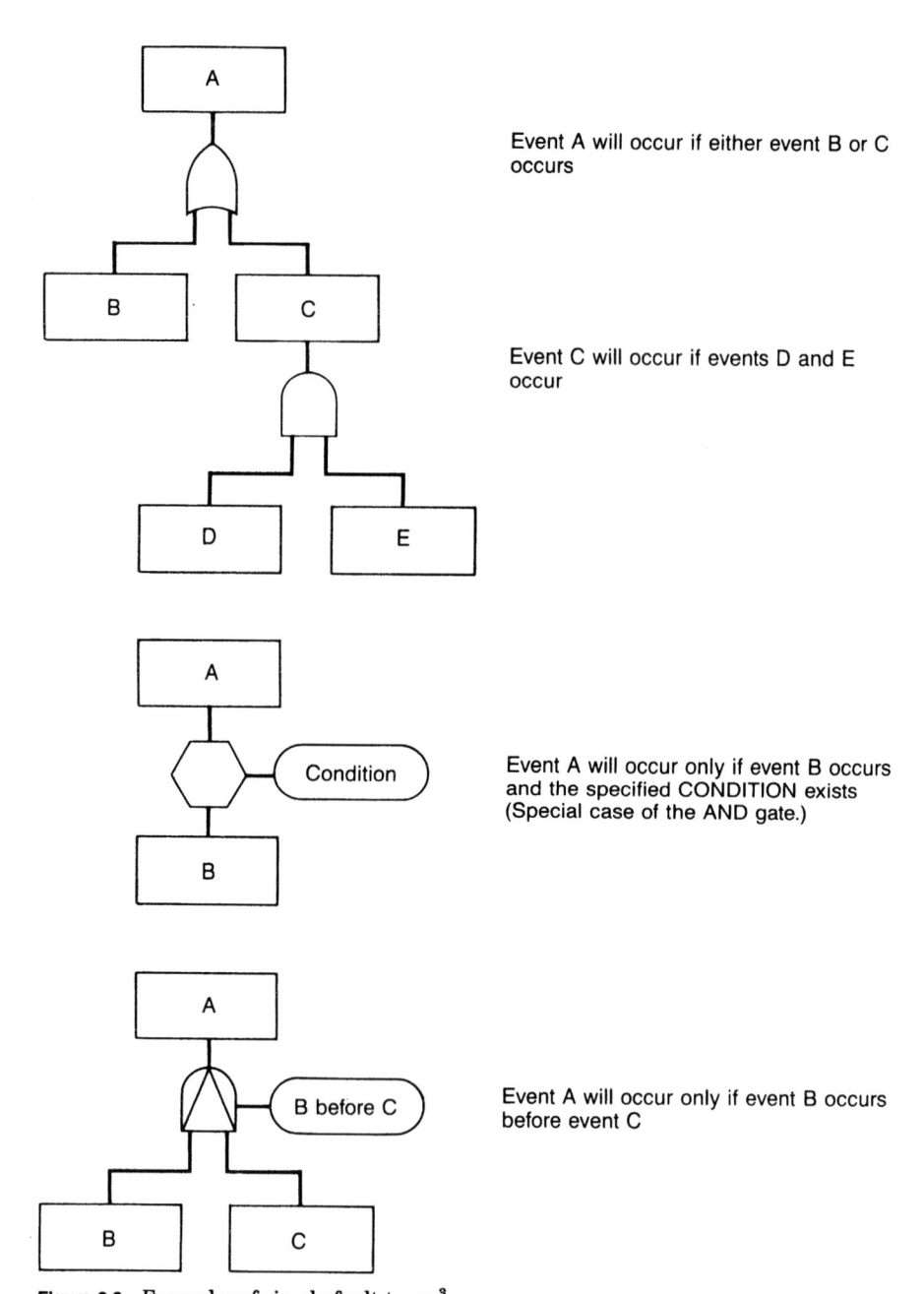

Event A will occur if either event B or C occurs

Event C will occur if events D and E occur

Event A will occur only if event B occurs and the specified CONDITION exists (Special case of the AND gate.)

Event A will occur only if event B occurs before event C

Figure 9.2 Examples of simple fault trees.[3]

Hardware Assessment and Validation 161

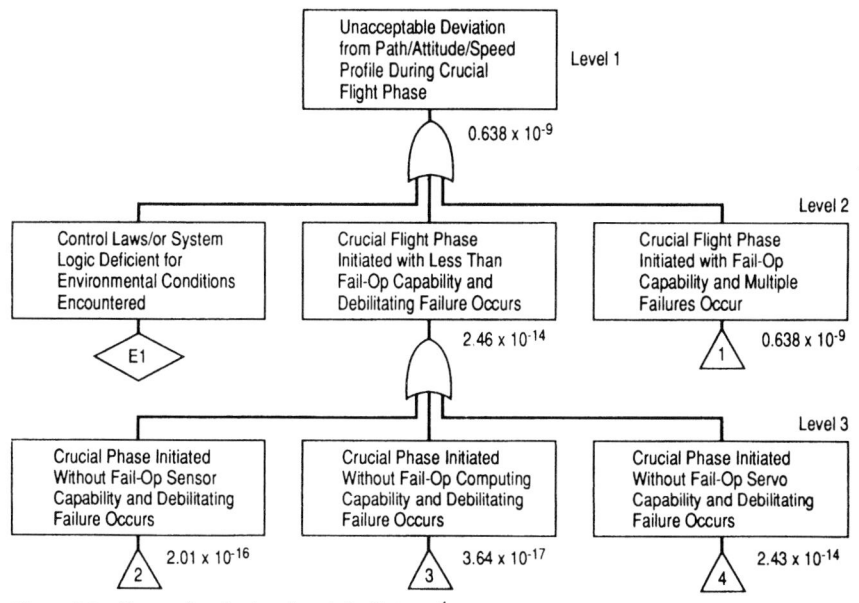

Figure 9.3 Example of a top-level fault tree.[4]

Fault tree diagram:

- **Level 1**: Unacceptable Deviation from Path/Attitude/Speed Profile During Crucial Flight Phase — 0.638×10^{-9}

- **Level 2**:
  - Control Laws/or System Logic Deficient for Environmental Conditions Encountered (E1)
  - Crucial Flight Phase Initiated with Less Than Fail-Op Capability and Debilitating Failure Occurs — 2.46×10^{-14}
  - Crucial Flight Phase Initiated with Fail-Op Capability and Multiple Failures Occur — 0.638×10^{-9} (Triangle 1)

- **Level 3**:
  - Crucial Phase Initiated Without Fail-Op Sensor Capability and Debilitating Failure Occurs — 2.01×10^{-16} (Triangle 2)
  - Crucial Phase Initiated Without Fail-Op Computing Capability and Debilitating Failure Occurs — 3.64×10^{-17} (Triangle 3)
  - Crucial Phase Initiated Without Fail-Op Servo Capability and Debilitating Failure Occurs — 2.43×10^{-14} (Triangle 4)

of the control laws being deficient for the environmental conditions encountered is assumed to have a zero probability; that is, for the purposes of the FTA, the control laws are considered to be perfect. Therefore, the event is not developed further. (Chap. 10 discusses software validation.) The center Level 2 event of crucial flight phase initiated with less than fail-op capability and debilitating failure occurs has a probability of 2.46×10^{-14} as determined by the Level 3 events. Like the Level 1 event, any one of three Level 3 events could cause the center Level 2 event to happen. The right Level 2 event of crucial flight phase initiated with fail-op capability and multiple failures occur has a probability of 0.638×10^{-9} and is the major determinant of the probability of the Level 1 event. Triangle 1 under the right Level 2 event means that the supporting Level 3 fault tree(s) are continued on another sheet.

Since fault trees deal with events and relationships, it is natural that boolean algebra, the algebra of events, would play a major role in fault tree analysis. Fault trees are graphical portrayal of boolean equations which means that, conversely, the trees can be reduced to an equivalent set of boolean equations. This tree-equation equivalence is a very powerful relationship that can be exploited in the automated analysis of trees. The reader is advised to examine a text on boolean algebra or the topic discussion in NUREG 0492 before attempting an analysis of a tree of more than modest complexity.

An important concept in FTA is minimum cut sets. A minimum cut set is a smallest combination of events which, if they all occur (under prescribed conditions or sequences, if any), will cause the top event to happen. In most trees there are at least several minimum cut sets. A one-component minimum cut set represents a single failure that will cause the top event to happen. The determination of minimum cut sets is best accomplished by the conversion of the tree to equivalent boolean equations and then applying the laws of boolean algebra to simplify and reduce the tree to minimum cut sets. Boolean algebra is a rigorous technique that is required to ensure that all possible cut sets are identified.

FTA has received wide acceptance because of the important results that it can produce. Results from FTA can be either qualitative or quantitative. Qualitative results include minimum cut sets discussed above, qualitative importance of a function or component, and common cause potentials. Quantitative results include numerical probabilities, quantitative importances, and sensitivity evaluations.

Qualitative importance refers to the number of elements in the minimum cut sets. Sets containing only one element, representing single-point failures, are judged to be the most important and demand more attention than those containing two elements, which, in turn, demand more attention than those containing three elements, and so on. Experience has shown that an examination of cut sets containing more than three elements is rarely required since the probability of that cut set occurring is extremely low compared to cut sets containing fewer elements.

Minimum cut sets can also be examined to identify common causes of failure in a complex system. Cut sets ease the identification of elements subject to malfunction or failure by a single common cause. Typical common causes of failure in avionics equipment include loss of power, electromagnetic interference, and breakdown of data buses.

Once minimum cut sets are available, quantitative probability calculations can be made and the contribution of each component to the unreliability of the system can be determined. The contribution can be determined for each component in a given cut set or for each component to the total unreliability of the system when all cut sets containing that component are summed.

Sensitivity studies can be easily conducted on minimum cut sets to answer a range of "what if" questions such as the effect of changing the inspection period, changing to a component with different reliability, and so on.

Failure Modes and Effects Analysis

FMEA complements FTA. While FTA addresses the top-level questions (for example, given a failure of the system, what are the possible causes?), FMEA takes the opposite approach. It is inductive, going from the specific to the general. FTA can be productively applied down to approximately the circuit board level in a system design, beyond which it becomes hopelessly complex and voluminous. Thus, the approach is reversed at that point and the bottom-up FMEA procedure, is applied.

FMEA identifies all possible failure modes, typically at the pin or individual component level, and determines the immediate board-level effects. The board-level effects become the lowest resolvable elements in a FTA. If desired, it is possible to continue the FMEA process to the top level of the system in lieu of the FTA process.

FMEA requires cataloging and listing of all possible failure modes of every component (including connectors and wires) followed by a thorough examination of each failure mode to determine its impacts on design and system operation and the classification of the severity of these impacts.

The material in this section is based primarily on MIL-STD-1629 Procedures for Performing a Failure Mode, and Effects and Criticality Analysis.[6] MIL-STD-1629 is intended for military use, but the techniques suggested can be applied to FMEAs prepared in response to the FAA Advisory Circular 25.1309-1A.[2]

For maximum effectiveness, the FMEA process should begin as early as possible in the design process to permit the systematic classification of design shortcomings and performance constraints and should continue in an iterative mode throughout the design and testing of the hardware. FMEA applies only to hardware; it is not adaptable to software.

The basic steps in an FMEA according to 1629A are as follows:

1. Define the system to be analyzed.

2. Construct block diagrams.

3. Identify all potential items and interface failure modes and define their effects at all levels up to and including the mission.

4. Evaluate the consequences of each failure mode and assign a severity classification.

5. For each failure mode identify detection methods and compensation provisions.

6. Identify design changes or corrective actions to eliminate or contain the failure.

7. Identify effects of actions generated in step 6 above on other system attributes.

8. Document uncorrectable problems and specify actions necessary to reduce the risk.

To support the process in step 4 above, the following criticality classifications are adopted from MIL-STD-882: System Safety Program Requirements:[7]

Class I: Catastrophic failure effect: Results in the immediate forced landing of the aircraft, aborted mission, and probable damage and injury to the aircraft and personnel, respectively.

Class II: Critical effect: Allows recovery to forward flight and subsequent return to the launch base under severely degraded flight control. Mission is aborted and there is possible damage and injury.

Class III: Slight effect: Provides mission completion capability with slight degradation in flight control. Low probability of damage and injury.

Class IV: No effect: Requires unscheduled maintenance.

An FMEA assumes a single failure within a clock cycle of the unit under analysis. Multiple failures, even within a few consecutive clock cycles, would be too complex to analyze. Another general rule is that an FMEA is performed for each replaceable module in the system, and in these cases, the module is treated as a system.

Table 9.2 shows the first step in an FMEA. In this example a microprocessor from a hypothetical flight control computer is being analyzed on a pin-by-pin basis. Note that each pin, or set of common pins, is identified, the type of fault is stated [open, ground (short), or high], and the effect on the microprocessor performance is then stated in very precise terms. If a failure effect is ambiguous or indeterminate, fault injection testing may be necessary. *All* predictable failure modes must be identified. This is the single most important (and most demanding and onerous) task in the FMEA process. No potential failure mode can be omitted. This information is then adapted to the FMEA worksheet.[4]

An FMEA worksheet is shown in Fig. 9.4[6]. Indenture level refers to the level of analysis, which can range from the entire aircraft to a single electrical component. The type of mission must be specified since the effects of a given failure may vary depending on the mission of the aircraft. Identification number is a reference designation for traceability purposes. (These numbers are usually assigned in accordance with MIL-STD-780: Work Unit Codes for Aeronautical Equipment; Uniform Numbering System.)

Mission Phase/Operational Mode is a more concise and specific statement of what the aircraft is doing at the time of the postulated failure than is given in the heading of the form. This information is important in establishing the precise end effects of the failure.

TABLE 9.2 Example Failure Modes and Effects Analysis for a Typical Microprocessor[4]

Circuit	Function	Pin	Fault	Effect
Microprocessor DU 17	Process the low-order bits of the 16-bit CAPS word in response to instructions from the control registers.	CN	Open	Microprogram address is not incremented during execution of microcode sequence. Massive processor failure.
			Ground	Same as above.
		A0–A3	Open	Wrong A pointer address when failed bit should be high. Massive processor failure.
			Ground	Same as above.
			High	Wrong A pointer address when failed bit should be low. Massive processor failure.
		B0–B3	Open	Wrong B pointer address when failed bit should be high. Massive processor failure.
			Ground	Same as above.
			High	Wrong B pointer address when failed bit should be low. Massive processor failure.
		I0–I2	Open	Wrong data source selected when failed bit should be high. Massive processor failure.
			Ground	Same as above.
			High	Wrong data source selected when failed bit should be low. Massive processor failure.
		I3–I5	Open	Wrong operation performed when failed bit should be high. Massive processor failure.
			Ground	Same as above.
			High	Wrong operation performed when failed bit should be low. Massive processor failure.
		I6–I8	Open	Wrong destination code when failed bit should be high. In most cases, the immediate effect will be internal to the chip involving load or shift of data in registers. Massive processor failure.
			Ground	Same as above.
			High	Wrong destination code when failed bit should be low. Massive processor failure.

TABLE 9.2 Example Failure Modes and Effects Analysis for a Typical Microprocessor[4] (Continued)

Circuit	Function	Pin	Fault	Effect
		CP	Any	Chip dead. Massive processor failure.
		D0–D3	Open	Input to processor is wrong when failed bit should be high. Major effect caused by incorrect bit in packed boolean data. Massive processor failure.
			Ground	Same as above.
		C	Open	Carry-in always low. Program counter not incremented on instruction fetch. Massive processor failure.
			Ground	Same as above.
			High	Carry-in always high. Foreground loop of flight software cannot execute paths 2 and 4; iteration monitor test bit not toggled; iteration monitor trips. Flight control computer (FCC) disconnects.
		Y0–Y3	Open	Wrong address gated on CAPS address line. Massive processor failure.
		P	Open	Carry propagate always sent to carry look-ahead logic. Massive processor failure.
			Ground	Same as above.
			High	Carry propagate never sent to carry look-ahead logic. Double-precision integrators drift.
		G	Open	Carry generate always sent to carry look-ahead logic. Massive processor failure.
			Ground	Same as above.
			High	Carry generate never sent to carry look-ahead logic.
		F = 0	Open	DU17 cannot pull down F = 0 line to status register, yielding false results for some logic tests. Massive processor failure.
			Ground	DU17 always pulls down F = 0 line to status register, yielding false results for some logic tests. Massive processor failure.
			High	Same as open.

V_{cc}	Open	Chip dead. Massive processor failure.
OE	Open	Chip dead. Massive processor failure.
Ground	Open	Chip dead. Massive processor failure.
R3	Open	Bit left shifted into DU17 or right shifted into DU14 always low. Multiplication results erroneous. FCC disconnect.
	Ground	Same as above.
	High	Same as high.
R0	Open	Bit right shifted to shift/rotate multiplexer or input from shift/rotate multiplexer always low. Multiplication results erroneous. FCC disconnects.
	Ground	Same as above.
	High	Bit right shifted to shift/rotate multiplexer or input from shift/rotate multiplexer always high.
Q3	Open	Bit right shifted into DU17 or left shifted into DU14 always low. Multiplication results erroneous. FCC disconnect.
	Ground	Same as above.
	High	Bit right shifted into DU17 or left shifted into DU14 always high. Multiplication results erroneous. FCC disconnect.
Q0	Open	Bit right shifted to shift/rotate multiplexer or left shifted from shift/rotate multiplexer always low. Multiplication results erroneous. FCC disconnect.
	Open	Bit right shifted to shift/rotate multiplexer or left shifted from shift/rotate multiplexer always high. Multiplication results erroneous. FCC disconnect.
F3	Any	No effect. Pin not connected.
CN + 4	Any	No effect. Pin not connected.
OVR	Any	No effect. Pin not connected.

FAILURE MODE AND EFFECTS ANALYSIS

SYSTEM _____
INDENTURE LEVEL _____
REFERENCE DRAWING _____
MISSION _____

DATE _____
SHEET _____ OF _____
COMPILED BY _____
APPROVED BY _____

Identification Number	Item/Functional Identification (Nomenclature)	Function	Failure Modes and Causes	Mission Phase/ Operational Mode	Failure Effects			Failure Detection Method	Compensating Provisions	Severity Class	Remarks
					Local Effects	Next Higher Level	End Effects				

Figure 9.4 Example of a FMEA worksheet format.

Failure effects are as important as failure mode, and it requires expert knowledge of the system to identify all of the effects. Failure effects are classified as local, next higher level, and end effect. Local refers to effects on the unit under analysis, the next higher level is self-explanatory, and end effect refers to the impact on the top-level system; that is, a jammed control surface (local effect) can result in the loss of the aircraft (end effect). The failure detection method describes how the failure can be detected (if possible) by the crew through instrument indicators, warning devices, and so on. If the failure cannot be detected, potential subsequent failures need to be examined to determine if they will be evident to the crew. Methods of isolating the failure are listed in this column also. Compensating provisions are design features or crew actions which circumvent or mitigate the effects of the failure. Examples of compensating provisions include redundancy and safety or relief devices. Severity classification refers to one of the four classes defined above, with Class I being the most severe and Class IV the least.

In adapting FMEA for civil applications, a new column could be substituted for the severity class column to address the question of whether or not the flight could be dispatched with the equipment inoperable and still meet the minimum equipment list (MEL) requirements.*

Criticality and Damage Modes and Effects Analyses

For military avionics, the FMEA is the precursor to two other important analyses: criticality analysis (CA) and damage modes and effects analysis (DMEA). These additional analyses are, in essence, more detailed examinations of the "effects" part of the FMEA.

In the CA, as described in MIL-STD-1629, the criticality of a failure mode is established by combining its severity classification, as determined in the FMEA, with an estimate of its probability of occurrence. Like the FMEA, there are two approaches to CA, qualitative and quantitative.

In the qualitative approach to CA, the probability of failure is estimated by the designer or analyst and ranges from 0.20 for "Level A—Frequent" failures to 0.001 for "Level E—Extremely Unlikely" failures, as shown in Table 9.3. (Note this definition of extremely

---

*The MEL is a carefully determined (based on aircraft design, route structure, and maintenance procedures), rigidly interpreted list of equipment, approved by the FAA, that a civil transport must have in operating condition before being permitted to depart on a flight. The MEL may change depending on the expected enroute weather and the level of maintenance at the destination airport.

TABLE 9.3 MIL-STD-1629 Criticality Analysis Probability of Failure

Level	Adjective description	Probability
A	Frequent	0.2
B	Reasonably probable	0.1 to 0.2
C	Occasional	0.01 to 0.1
D	Remote	0.001 to 0.01
E	Extremely improbable	0.001

improbable is substantially different from the definition of extremely improbable in FAA Advisory Circular 25.1309-1A.)

In the quantitative approach to CA, data from recognized failure rate sources (such as MIL-HDBK-217: Reliability Prediction of Electronic Equipment) are used, as modified by environmental and quality factors.

A CA worksheet is shown in Fig. 9.5. Note that the first six columns are data generated by the FMEA and simply transferred to this sheet. Failure effect probability β is the conditional probability that the failure will result in the designated criticality of failure. Estimated β values, assigned by the analyst, range from 1.0 for actual loss through probable and possible loss to 0 for no effect.

The failure ratio α is the fraction of the failure rate as stated in MIL-HDBK-217 or other reliability data source represented by the stated failure mode, e.g., bit hung high, bit hung low, and so forth. The sum of the α values is 1.0 for all of the possible failure modes of a given component. The component failure rate λ_p is extracted from MIL-HDBK-217 or is estimated by the analyst as discussed earlier. The operating time t is derived from the mission description. Once these data are in hand, the failure mode criticality C_m is calculated by the equation

$$C_m = \beta \alpha \lambda_p t$$

The failure mode criticality number represents the contribution to the item criticality number for one failure mode in a given severity classification. The item criticality number C_r is the sum of the failure mode criticality numbers for an item for a particular severity classification and mission phase. It is computed by the following formula:

$$C_r = \sum_{n=1}^{j} (\beta \alpha \lambda_p t)_n \quad n = 1, 2, 3, \ldots, j$$

The final step in the CA is the preparation of the criticality matrix, which compares each failure mode to all others in terms of severity.

CRITICALITY ANALYSIS

SYSTEM _____
INDENTURE LEVEL _____
REFERENCE DRAWING _____
MISSION _____

DATE _____
SHEET _____ OF _____
COMPILED BY _____
APPROVED BY _____

Identification Number	Item/Functional Identification (Nomenclature)	Function	Failure Modes and Causes	Mission Phase/ Operational Mode	Severity Class	Failure Probability Failure Rate Data Source	Failure Effect Probability (β)	Failure Mode Ratio (α)	Failure Rate (λ_p)	Operating Time (t)	Failure Mode Crit $C_m = \beta\alpha\lambda_p t$	Item Crit $C_r = \Sigma(C_m)$	Remarks

Figure 9.5 Example of CA worksheet format.

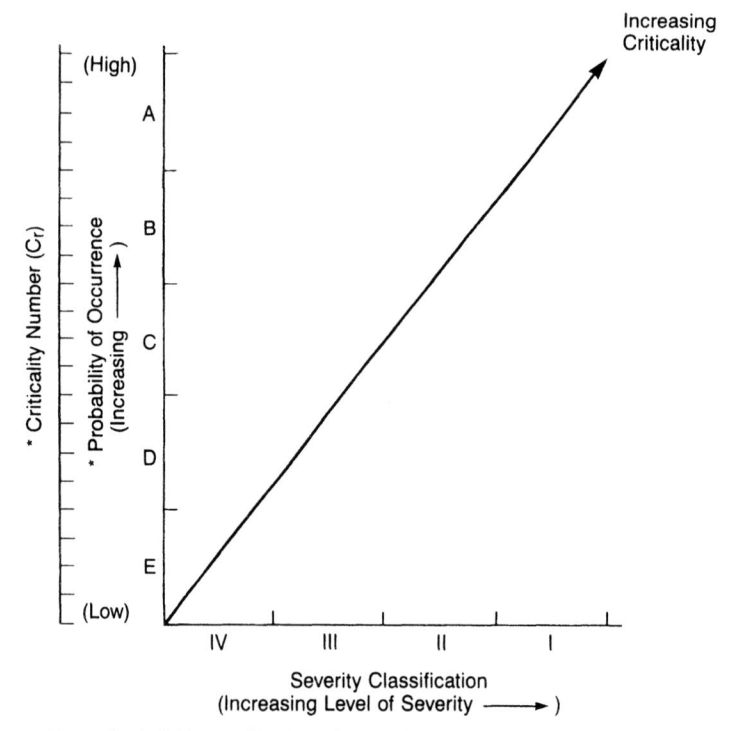

Figure 9.6 Example of criticality matrix.[6]

Shown in Fig. 9.6 is the form for a criticality matrix. The ordinate is either the criticality number C_r or the probability of occurrence level, as determined from the procedures described above. The abscissa is the assigned severity classification from the FMEA. In this plot of the failure mode data, those modes which require the most attention are further along the diagonal, near the upper right corner.

In addition to the CA required for military aircraft, a DMEA may also be required, especially for high-performance aircraft. The DMEA quantifies the survivability of aircraft systems, in this case the avionics. The procedures reviewed in this section can be found in MIL-STD-2069: Requirements for Aircraft Nonnuclear Survivability Program.

Survivability is defined as the capability of a system to continue to function in the presence of a nonnuclear threat. It is frequently expressed as a probability of a kill given a hit, $P(k/h)$, and is a function of susceptibility and vulnerability.

Susceptibility is a measure of the probability that an object will be hit by a given threat mechanism. It is derived through the consideration of the following characteristics:

Signatures

Mission profile

Countermeasures

Location in aircraft

Hardening

Vulnerability is a measure of characteristics that contribute to the degradation or loss of function in an object. Critical characteristics in assessing vulnerability include:

Critical areas

Fault tolerance

Redundancy

Secondary potential failure

The DMEA builds on the FMEA data, associates each failure mode of the FMEA with threat weapons and damage mechanisms and quantifies the response of the unit under analysis to the damage mechanism. There are four outputs from the DMEA process: (1) DMEA matrix, (2) disablement diagrams, (3) fault trees (not to be confused with the fault trees discussed earlier in this chapter), and (4) $P(k/h)$ functions.

Two of these outputs deserve special mention. The disablement diagram, similar in appearance to an annotated isometric drawing of a major subsystem, contains four items: component and/or subsystem relative physical location, statements on how these items can fail, effects of failure on higher systems and the aircraft, and the results of these failures expressed in the kill criteria of the aircraft.

The $P(k/h)$ plot is the second output that should be noted. A plot of $P(k/h)$ versus fragment velocity for two designs of F-14 rudder actuators, when exposed to a 12.7-mm armor-piercing incendiary round, is shown in Fig. 9.7. Susceptibility is represented by the distance from the origin to the beginning of the curve—the greater the distance, the lesser the susceptibility. Vulnerability is represented by the final height of the curve—the higher the curve, the greater the vulnerability. Thus, in the curves shown, both actuators have approximately the same susceptibility but the electromechanical actuator RE-300 is less vulnerable.[8]

DO-160 Test Requirements

Hardware assessment and validation are more than just analytical techniques and modeling the system in a computer. Once the LRU or line replaceable module (LRM) boxes have been assembled and given

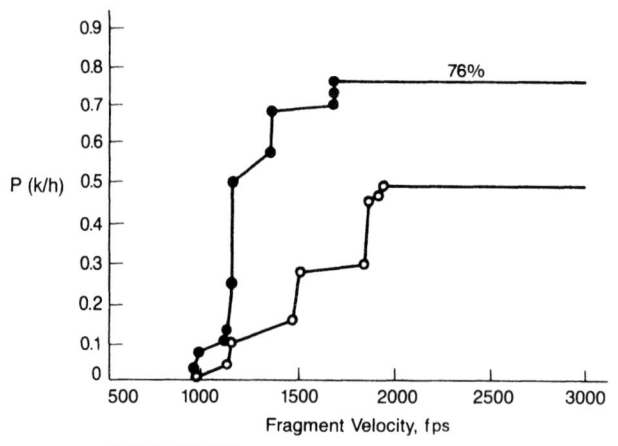

Figure 9.7 P (k/h) curves for candidate F-14 rudder actuators—12.7-mm armor piercing incendiary threat.[8] (© *American Institute of Aeronautics and Astronautics; reprinted with permission.*)

a bench check, the next step is to subject them to a series of environments and tests that closely match the conditions found in operational service. Avionics destined for civil aircraft are tested in accordance with RTCA document DO-160: Environmental Conditions and Test Procedures for Airborne Equipment.[9] FAA Advisory Circular 21-16C states that the procedures in DO-160 may be used in Technical Standard Order (TSO) authorizations.[10]

DO-160 describes 21 tests that avionics may be subjected to in determining acceptability for service in civil transports. The document describes the test, test equipment, and procedures. The purpose of each test, as well as additional information on several important tests, have been extracted from DO-160 and are included here in Part A of App. B. Most of the tests require that the equipment being tested be designated by different categories in which the equipment must operate to reflect the severity of the environment and, consequently, the test. In many cases the categories simply designate whether the equipment is subject to the particular environment or not. Since some of the tests could be potentially damaging to the equipment, the order in which the tests are performed must be carefully determined.

The first five tests described in the document include (1) temperature and altitude, (2) temperature variation, (3) humidity, (4) operational shocks and crash safety, and (5) vibration. In the temperature and altitude test, equipment may be subject to reduced pressures equivalent to an altitude of 70,000 ft and ambient temperatures from −55°C (−67°F) to 85°C (+185°F) depending on the type of aircraft and where the equipment will be mounted. For the vibration tests, a series of spectra are given for various combinations of aircraft and propulsion modes.

Three tests examine the immunity of the equipment to liquids and salt spray. One test is devoted exclusively to determining if the equipment is waterproof. Water ranging from dripping condensation to a high-pressure stream is used to test the equipment. Another test determines the equipment susceptibility to common aircraft fluids including fuel, hydraulic fluids, lubricating oils, solvents and cleaning fluids, de-icing fluids, fire extinguishants, insecticides, and sullage.

Equipment may also be tested to determine its susceptibility to sand and dust, ice, and its resistance to fungus. There is also a test to establish the influence of the equipment on the magnetic indicators in the cockpit.

Of singular importance in commercial passenger aircraft is the explosionproofness test. The test demonstrates that the equipment is explosionproof if the equipment, under normal conditions, comes in contact with flammable fluids or vapors. Equipment that may come in contact with flammable fluids and vapors in "non-normal" conditions must be shown to be designed and installed to minimize the risk of fire and/or explosion.

Of particular interest to the avionics designer is a series of six tests to determine equipment performance in an array of electrical environments. The first two tests establish the equipment performance in the presence of off-nominal steady state voltages, power interruptions and surges, and spikes on the input power lines. These tests are described in detail in the second section of Chap. 4.

The last group of tests described in DO-160 can be grouped under the general heading of electromagnetic interference (EMI). The frequencies of interest range from audio through radio frequency (RF). The questions of radiating power as well as susceptibility to power in these frequencies are examined. Details on these tests are summarized later in this chapter.

Throughout the test procedures, the statement "Determine compliance with applicable equipment standards" is frequently made. The performance standards referred to are the RTCA minimum performance standards (MPS) and/or RTCA minimum operational performance standards (MOPS) or the European equivalents EUROCAE MPS and EUROCAE MOPS for that particular type of equipment, whether it is a radio or an autopilot.

The manufacturer must prepare an Environmental Qualification Form as part of the data package submitted for the TSO authorization for the equipment. This form should follow the format shown in DO-160 and describe each environment (with category, where applicable) to which the equipment was subjected.

As thorough as DO-160 is, there are still additional environmental tests that may be required by the airframe manufacturer or by the end user but are not included in the document. Examples of tests not covered include hail, acoustic, vibration, and combined temperature, humidity, and pressure tests.

MIL-STD-810 Test Requirements

Military avionics are subject to environmental tests that are similar in design and intent to those for civil avionics. MIL-STD-810 Environmental Test Methods and Engineering Guidelines governs the environmental testing of all military hardware including avionics. MIL-STD-810 is much broader in scope than RTCA document DO-160 and addresses environmental tests for the equipment under *all* conditions including shipping, storage, and operation.[11] The purpose of each test in MIL-STD-810 is stated here in Part B of App. B.

In contrast to many military standards, 810 encourages flexibility and judgment in interpreting the requirements set forth therein. Much emphasis is placed on "tailoring" the tests to the unique needs of the equipment under test. The definition of tailoring given in the standard reads, in part, "the process of choosing or altering test procedures, conditions, values, tolerances, measures of failure, etc., to simulate or exaggerate the effects of one or more forcing functions." It is recommended in the standard that an environmental engineering specialist select and tailor the standard tests to each particular case.

As a way of reducing the amount of environmental testing required, 810 recognizes field and fleet data as suitable substitutes for test data if certain conditions are met. The equipment in the field must be similar to the equipment under test for those parameters being tested. However, the field equipment does not have to be functionally equivalent. Also, the field data must meet rigorous quantity and quality requirements.

As noted earlier, the complete set of 20 tests described in 810 is very similar in intent and procedures to those in DO-160; however, there are some notable exceptions. There are the traditional tests such as low pressure, high and low temperature, sand and dust, and explosive atmosphere. Other specialized tests which focus on military needs include gunfire and acoustic noise.

The Temperature, Humidity, Vibration, Altitude Test (Method 520.0) is the major test for avionics that assesses the combined effects of these environmental factors. It is intended for avionics mounted inside an aircraft as opposed to externally carried sensors and stores. There are three test procedures described, each intended for use in different phases of avionics systems development.

The first test procedure is the *Engineering Development Test*. The emphasis is on finding defects early in the design cycle when they are relatively easy and inexpensive to correct. This test is frequently accelerated by eliminating benign conditions and using higher than normal stresses. During the test, the test item should experience the equivalent of 300 to 600 mission hours. If few or no failures occur, the test can be terminated early.

The second test procedure is the *Flight or Operations Support Test*. The goal of this procedure is to minimize the costly and complex flight test requirements by simulating flight conditions on the ground. This test should be run as long as required to show that environmental conditions will not cause problems.

The third and most important test is the *Qualification Test*. This is a formal, rigorously controlled and conducted test to demonstrate compliance with contract requirements. It emphasizes the most significant environmental stresses and includes the maximum amplitude of each stress and significant unique but plausible combinations of stresses. Typically the test cycle is repeated 10 times.

For the purposes of 810 testing, five types of missions have been identified:

Bombing and combat training

Radar training

Operational training

Transition training

Function checkout

Each mission is broken into phases like those discussed in Chap. 1. For each mission, avionics bay temperature, humidity, pressure, and vibration profiles, such as those shown in Fig. 9.8 are generated. Two test cycles should be developed, one to simulate routine usage and one to simulate combat or combat training conditions.

MIL-STD-810 includes some very modest electrical stress testing. For example, ac and dc input voltages should be varied from test cycle to test cycle from 90 to 110 percent of nominal. The equipment under test should be switched on and off, and the power system output should be varied to simulate temporary system overload and regulating devices fluctuations.

When compared to DO-160, MIL-STD-810 does not include some very important tests. There are no input voltage tests since these are covered by the requirements in MIL-STD-704: Aircraft Electrical Power Characteristics.[12] Also, there are no EMI tests since they are covered in MIL-STD-461: Electromagnetic Emission and Susceptibility Require-

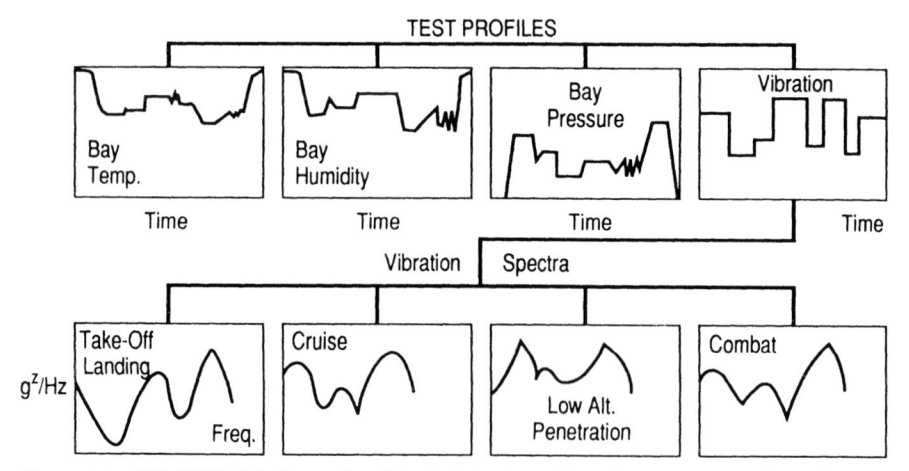

Figure 9.8 MIL-STD-810 Example of typical test profiles of temperature, humidity, pressure, and vibration.

ments for the Control of Electromagnetic Interference[13] and MIL-E-6051 Electromagnetic Compatibility Requirements, Systems.[14] (See Chap. 4 for a discussion of power requirements and the next section for a discussion of EMI.)

Electromagnetic Interference

EMI is a major problem in modern digital avionics systems, and it is getting worse. This situation is the result of smaller circuit elements that require less energy to be damaged or change their state, installation of avionics in aircraft made of composite material that offers substantially less shielding to external EMI fields, and more and higher-powered man-made EMI sources (radio and television stations, etc.). Man-made sources are usually called high intensity RF (HIRF). The most common natural EMI source is lightning.

One of the most vexing aspects of EMI is that it is somewhat aircraft dependent, so some problems may not appear until the avionics are installed in the aircraft. EMI is very difficult to troubleshoot and will tax the abilities of even the best designers. This section overviews the requirements for reducing the effects of EMI, describes tests to demonstrate meeting these requirements, and offers proven design techniques for achieving a sound, EMI-tolerant design.

Figure 9.9 shows the internal and external sources of EMI for a typical aircraft. There are three basic sources of EMI that have been further divided into ten specific sources. Electrostatic discharge is the

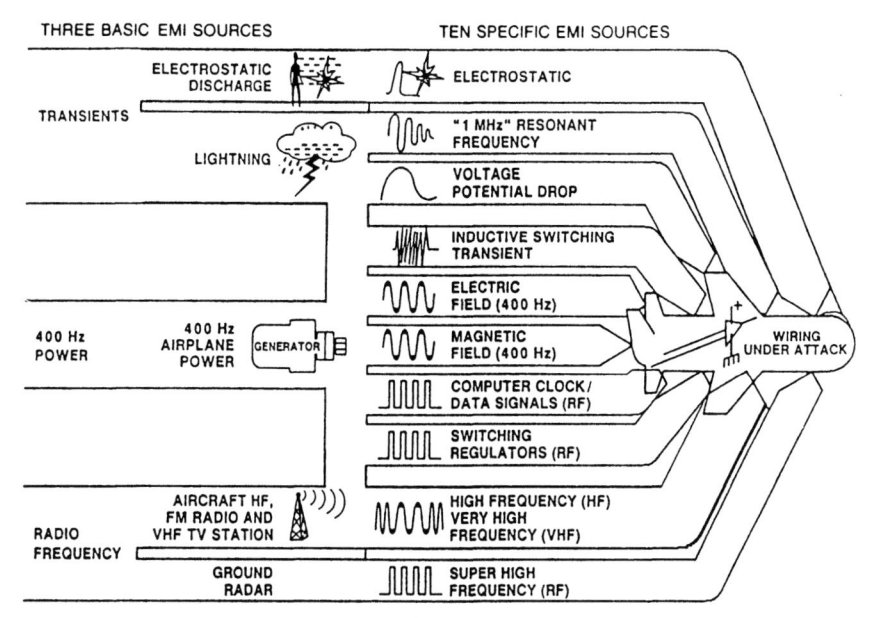

Figure 9.9 Representative EMI sources.[15]

charge that builds up on an aircraft from simply moving through the air. There are two types of EMI from lightning strikes to the aircraft: an approximately 1-MHz (depending on the aircraft length) resonant current induced in the aircraft by the lightning current and a voltage drop along the path of the lightning current through the aircraft generated by the current flowing through the small but significant resistance of the aircraft. The onboard 400-Hz power system generates five forms of EMI, all of which are predictable and relatively easy to accommodate. The HIRF sources of EMI are divided roughly according to frequency. Not shown as HIRF sources are the industrial RF emitters used as a heat source for some production processes. Radar HIRF sources are of special concern because of the high-power density in the focused beam.[15] Figure 9.10 shows the frequency of EMI sources compared to avionics operating frequencies.

In dealing with EMI on military aircraft, the maze of requirements is almost as perplexing as the EMI problems. There are requirements at the system, subsystem, and LRU levels. The most important document in establishing overall EMI requirements for military avionics systems is specification MIL-E-6051: Electromagnetic Compatibility Requirements, Systems. The specification spells out the need for a system-level electromagnetic compatibility plan and a board to ad-

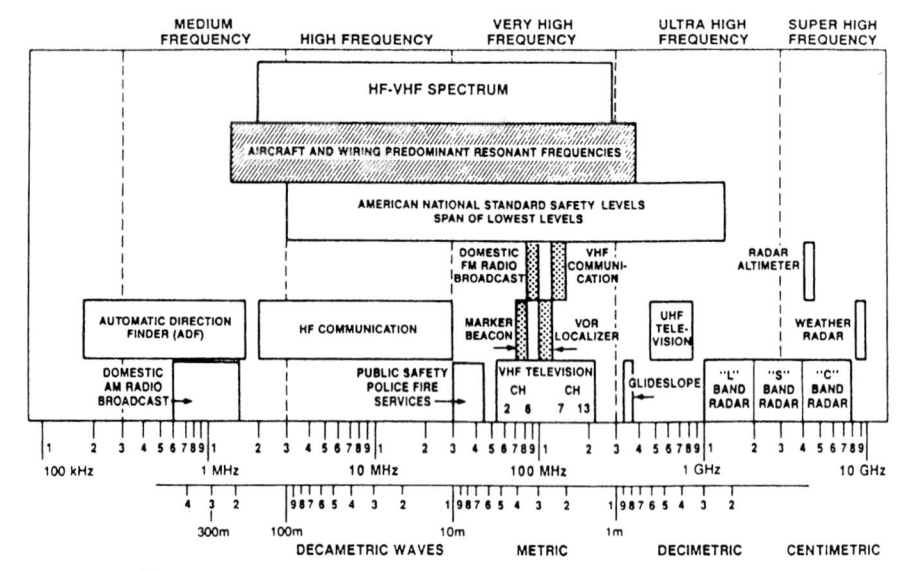

Figure 9.10 Frequency comparison: EMI sources versus avionics operating frequencies.[15]

minister the plan and establishes three subsystem and equipment criticality categories:

Category I: Electromagnetic compatibility (EMC) problems that could result in loss of life, loss of vehicle, mission abort, or unacceptable reduction in system effectiveness

Category II: EMC problems that could result in injury, damage to vehicle, or reduction in system effectiveness that would endanger mission success

Category III: EMC problems that could result in annoyance, minor discomfort, or no reduction in system effectiveness[16]

Another requirement in MIL-E-6051 deals with transients on avionics power supply lines. Transients of less than 50-μs duration on dc supply lines shall not exceed +50 or −100 percent of the nominal voltage. For ac supply lines the range is ±50 percent of the nominal voltage. (Other power system voltage quality requirements can be found in MIL-STD-704: Aircraft Electrical Power Characteristics, discussed in Chap. 4.)

When dealing with subsystems and equipment, MIL-STD-461: Electromagnetic Emission and Susceptibility Requirements for the Control of Electromagnetic Interference establishes the requirements for EMI control.[13] This standard is one of a troika of EMI military standards. The others are MIL-STD-462: Electromagnetic Interference Characteristics, Measurement of and MIL-STD-463: Definitions and

Systems of Units, Electromagnetic Interference and Electromagnetic Compatibility Technology.[17,18] MIL-STD-461 "establishes the documentation and design requirements for the control of the electromagnetic emission and susceptibility characteristics of electronic, electrical, and electromechanical equipment and subsystems...." In essence, 461 presents a series of tests to which avionics equipment must be subjected, depending on the application of the equipment. These tests are very similar in concept and approach to the broad environmental tests described in MIL-STD-810, as discussed earlier in this chapter. Also, like MIL-STD-810, 461 permits tailoring the prescribed tests to fit the particular under test.

A total of 21 tests are described in 461, each designed to probe a different facet of the response of the equipment under test to EMI. The 18 tests listed in Table 9.4 are applicable to digital avionics. MIL-STD-461 presents equipment performance requirements that are to be verified during the tests. The companion document, MIL-STD-462, spells out the detailed test procedure for each of the tests.

RTCA document DO-160: Environmental Conditions and Test Procedures for Airborne Equipment establishes six tests to determine the susceptibility of civil avionics to EMI. These tests, listed in Table 9.5, are very similar in intent and protocol to those listed in Table 9.4.

Lightning is the most widely known (and awesome) form of EMI, and it presents the largest challenge to the aircraft and the avionics

TABLE 9.4 MIL-STD-461 Tests Required for Digital Avionics[13]

CE01	Conducted Emissions, Power and Interconnecting Leads, Low Frequency (up to 15 kHz)
CE03	Conducted Emissions, Power and Interconnecting Leads, 0.015 to 50 MHz
CE06	Conducted Emissions, Antenna Terminals 10 kHz to 26 GHz
CE07	Conducted Emissions, Power Leads, Spikes, Time Domain
CS01	Conducted Susceptibility, Power Leads, 30 Hz to 50 kHz
CS02	Conducted Susceptibility, Power Leads, 0.05 to 400 MHz
CS03	Intermodulation, 15 kHz to 10 GHz
CS04	Rejection of Undesired Signals, 30 Hz to 20 GHz
CS05	Cross-modulation, 30 Hz to 20 GHz
CS06	Conducted Susceptibility, Spikes, Power Leads
CS07	Conducted Susceptibility, Squelch Circuits
CS09	Conducted Susceptibility, Structure (Common Mode) Current, 60 Hz to 100 kHz
RE01	Radiated Emissions, Magnetic Field, 0.03 to 50 kHz
RE02	Radiated Emissions, Electric Field, 14 kHz to 10 GHz
RE03	Radiated Emissions, Spurious and Harmonics, Radiated Technique
RS01	Radiated Susceptibility, Magnetic Field, 0.03 to 50 kHz
RS02	Radiated Susceptibility, Magnetic Induction Field, Spikes and Power Frequencies
RS03	Radiated Susceptibility, Electric Field, 14 kHz to 40 GHz

C = conducted, R = radiated, E = emission, S = susceptibility.

TABLE 9.5 RTCA DO-160 Tests Required for Digital Avionics[9]

Audio Frequency Conducted Susceptibility—Power Inputs
Induced Signal Susceptibility
Radio Frequency Susceptibility (Radiated and Conducted)
Emission of Radio Frequency Energy
Lightning Induced Transient Susceptibility
Lightning Direct Effects (under development)

designers. Testing for immunity to the effects of lightning is a very difficult task, especially if the airframe must be subjected to simulated lightning strikes. Table 9.6 summarizes the recommended waveforms to be applied to the airframe when full-scale testing is required. These waveforms have been adopted by the FAA in Advisory Circular 20-136: Protection of Aircraft Electrical/Electronic Systems Against the Indirect Effects of Lightning.[19]

In civil transport aircraft, the EMI requirements are, in essence, captured by FAR Part 25.581 Lightning Protection which requires that "the airplane must be protected against the catastrophic effects of lightning." Compliance with this requirement in metallic aircraft can be achieved by proper bonding of the components to the airframe and/or by designing the components so that a strike will not endanger the airplane. Compliance with this requirement in composite aircraft can be achieved by design or by incorporating a suitable means of diverting the lightning-induced current away from the sensitive components.

While the array of tests designed to establish the EMI immunity characteristics of avionics is extensive, the techniques to minimize the susceptibility of equipment to EMI are relatively few and all can be basically characterized as sound engineering design. Perhaps the best reference on design techniques for reducing EMI susceptibility is MIL-HDBK-253: Guidance for the Design and Test of Systems Protected Against the Effects of Electromagnetic Energy.[20]

TABLE 9.6 SAE AE4L Committee Recommended Waveforms for Lightning Effects Testing

	Return stroke*	Restrike†	Multiple burst‡
I_{peak} (A)	200,000	50,000	10,000
di/dt (max) (A/s)	1.0×10^{11}	0.7×10^{11}	N.A.
t_{peak} (µs)	6.3	3.2	0.1
$t_{50\%}$ (µs)	69.0	34.5	N.A.

*"Main" lightning stroke.
†Restrike current; up to 24 in a 2-s period.
‡Random pulses; 24 sets of 20 pulses in a 2-s period.

MIL-HDBK-253 asserts that the best way to reduce EMI effects is to begin with a design that recognizes their potential impact. It suggests a layering approach be used in the design of equipment to prevent the penetration of spurious external signals. The approach begins with external shielding, cable protection, and, finally, protection for the components and circuits. Five specific protection techniques are described in detail: shielding, bonding, filtering, grounding, and circuit design.

Shielding is designed to keep radiated power either inside the originating subassembly or away from the susceptible circuit. Shielding most commonly takes the form of braided wire coverings, screens over openings, and the equipment case. For braided cables, the amount of surface covered by braiding is most important. In order to save weight, the amount of braiding is minimized, although too little braiding results in lower shielding effectiveness. Thus, there is a trade-off to be made. In many cases, rather than leave the decision to the designer, the amount of shielding is specified (e.g., MIL-STD-1553 requires the shielding to cover 90 percent of the surface of cables used for data buses. Where screens cover an opening, the size of the mesh must be small enough to shield against the highest-frequency EMI anticipated.

Bonding is required to ensure a low-resistance path between two metal parts. Bonds can either direct metal-to-metal contact or indirect contact through the use of a conductive jumper. If used, a jumper should be as short as possible so it will not in itself become an antenna for more EMI. As a general rule, a good bond will have a dc resistance of between 0.25 and 0.50 mΩ. Military specification MIL-B-5087: Bonding, Electrical and Lightning Protection specifies a dc resistance of less than 0.1 Ω for prevention of electrical shock and presents drawings of acceptable techniques to achieve this value.[21]

Filtering can be applied at the LRU level to eliminate EMI. Filters cause insertion losses, so their use should be minimized. A proper EMC plan will recognize the need for systemwide control of EMI and thereby reduce the number of filters required.

Proper grounding is essential when dealing with very low signal levels. There should be separate ground wire runs for signal returns, signal shield returns, power systems returns, and case or chassis ground. These ground wire runs must be tied together at a single reference point to prevent ground loops. The airline operators demand that particular attention be given to grounding since faulty grounds have been a particularly troublesome problem, especially after the aircraft has been in service several years.

The final protection technique recommended by MIL-HDBK-253 is careful attention to the circuit design. Some suggested guidelines include the use of differential amplifiers for common mode cancellation, use of frequency modulation instead of amplitude modulation where possible, and use of field effect transistors instead of bipolar devices.

To reduce the potential for EMI, signal cables should be routed as deeply as possible in cable channels. Also, where composites are being used, cables should be as close as possible to any substantial metal structure being used in conjunction with the composites. Such a location will at least partially offset the loss of shielding offered by the traditional metal aircraft structure and skin.

Quantitatively Evaluating Systems Designs

As noted earlier, digital avionics systems that perform flight critical functions must demonstrate through analysis and test, in accordance with FAA Advisory Circular 25.1309-1A, a failure rate of less than 10^{-9} per flight hour. The previous sections of this chapter have discussed FTA and FMEA as methods of demonstrating achievement of the required reliability goal. However, the very complex, highly redundant modern digital avionics systems virtually demand the use of computer-based reliability modeling and prediction.

Over the last decade, a number of computer-based reliability prediction models with a wide range of capabilities have been developed. Two of these models that are representative of the capabilities are Computer Aided Reliability Estimation, Version III (CARE III) and Semi-Markov Unreliability Range Estimator (SURE).

There are several caveats about using computer-based reliability prediction models. First, none has been formally validated. However, they have been used in a variety of applications, so they can be used in new situations with confidence. Second, these models address only hardware operational faults. They do not address design errors or software faults. Third, in selecting a model, the prospective user should review its capabilities and limitations and review the documentation to determine that it is adequate and understandable. Finally, the software must be written in a language compatible with the user's host machine.

CARE III is one of the best known and most widely used of the computer-aided reliability prediction tools. It has extensive capabilities and has been thoroughly tested and demonstrated through application to several advanced fault-tolerant computer designs. Another advantage of CARE III is the availability of the program through the Computer Software Management and Information Center (COSMIC).* A well-written *User's Guide* with example problems is also available.[22]

CARE III begins with a top-down description of the system. The hierarchical structure comprises, in descending order, the system, stage, and

---

*COSMIC, Computer Services Annex, University of Georgia, Athens, GA 30602.

module. The system is modeled using the FTA technique described earlier and can have up to 70 stages. Both temporal redundancy ("roll back" or "roll forward" techniques) and spatial redundancy (the use of spare modules) can be modeled. An intrinsic limitation of CARE III is that it cannot accommodate system dynamic reconfiguration and reallocation of resources. Markov modeling, as found in SURE, is required to properly present this facet of fault-tolerant systems.

CARE III can accommodate a wide range of fault parameters. Faults can be transient, intermittent, or permanent. The fault occurrence rate can be variable, as described by the Weibull function, or constant. Faults can be treated as active, benign, detected, error, or failure.

The strongest feature of CARE III is its ability to handle interdependencies and critically coupled faults. Critically coupled faults is the situation where a nonthreatening fault occurs in a stage or module at the same time as another unrelated nonthreatening fault is occurring in another stage or module. The effects of the faults combine to produce system failure. An example of critically coupled faults is a fault in a computer module and a simultaneous fault in a bus not tied to that computer. The faults combine to produce an unrecoverable failure condition where a receiver is tied to the failed bus and two other buses, one of which is connected to the faulted computer.

The output of the CARE III program is the probability of system failure at a selected mission time. This probability of failure is further broken down into (1) failure due to imperfect fault handling and (2) failure due to resource (e.g., module) depletion.

SURE is a newer (compared to CARE III) example of a computer-aided reliability prediction program. A fundamental characteristic of a fault-tolerant system is recovery from faults through reconfiguration and/or activation of spares. The developers of SURE recognized that in a fault-tolerant system the time between fault occurrences (triggered by failures), a "slow process," and the time required for fault recovery, a "fast process," are many orders of magnitude different and require different modeling approaches.*[23]

SURE models both the slow and the fast processes, fault occurrence and fault recovery, respectively. As noted above, the fault occurrence can be modeled using conventional component failure rate informa-

---

*Semi-Markov refers to modeling the behavior of a fault-tolerant system. A system that allows the traditional constant failure rate, like the rates found in MIL-HDBK-217, can be described in a conventional Markov model. However, in a fault-tolerant system which recovers by reconfiguration and spare elements being brought on line, the recovery transition times are substantially smaller than the fault occurrence intervals and do not follow a well-behaved pattern like the component failure rate.[24] Consequently, unique expressions must be developed for the recovery transitions. When the constant failure rates and the recovery transition rates are combined, the result is a semi-Markov model.

tion. The fault recovery modeling must be based on the results of fault injection experiments and/or on an analysis of the system design.

Figure 9.11 is the semi-Markov model of a triad computer system with one spare. Each circle represents a possible state of the system. The horizontal transitions are fault occurrences that occur at some basic failure rate λ times the number of processors operating. Vertical transitions are recovery processes. $F_1(t)$ is the model of the recovery time for activating the spare processor. $F_2(t)$ is the time to degrade to a simplex state. States 3, 6, 7, and 8 are "death" states.

Figure 9.12 is the narrative input model that defines the system shown in Fig. 9.11. LAMBDA is the failure rate of the processor. Since it is a slow process, no standard deviation information is required. MU1 is the mean time to replace a faulty processor (a fast process), and SIGMA1 is the associated standard deviation. Similarly, MU2 and SIGMA2 are the time and its associated standard deviation to degrade to a simplex. The bottom part of Fig. 9.12 shows the rate of transition from one state to another.

It is readily apparent from Fig. 9.12 that the narrative input model of any but the most simple systems would be extremely lengthy, complex, and possibly prone to errors. Thus, the Abstract Semi-Markov Specification Interface to the SURE Tool (ASSIST) was developed to answer this compelling need and to enhance the accuracy of the input model. ASSIST requires the user to enter the rules to describe the fault and recovery behavior of the system, and then it uses these rules to develop the input model. Since the rules are usually far fewer and

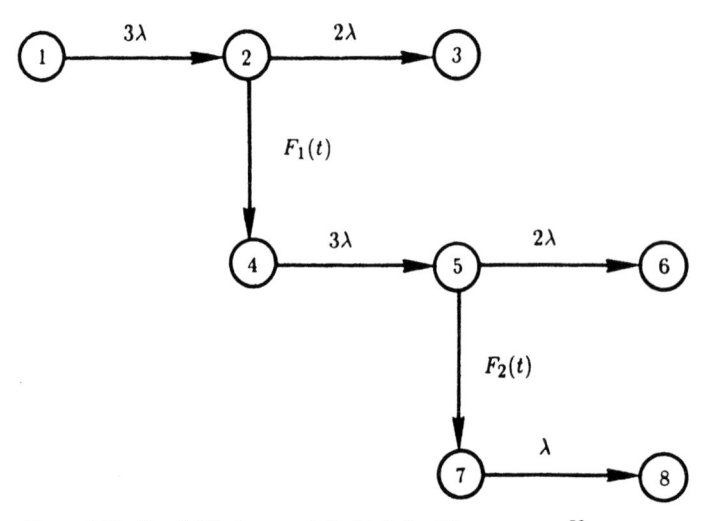

Figure 9.11 Semi-Markov model of triad with one spare.[23]

```
LAMBDA = 1E-4;          (* Failure rate of a processor *)
MU1 = 2.7E-4;           (* Mean time to replace faulty processor
                           w/ a spare *)
SIGMA1 = 1.4E-4;        (* Standard deviation of time to replace
                           w/ a spare *)
MU2 = 9.2E-4;           (* Mean time to degrade to a simplex *)
SIGMA2 = 3.8E-4         (* Standard deviation of time to degrade
                           to simplex *)
1,2 = 3*LAMBDA;
2,3 = 2*LAMBDA;
2,4 = <MU1, SIGMA1>;
4,5 = 3*LAMBDA;
5,6 = 2*LAMBDA;
5,7 = <MU2, SIGMA2>;
7,8 = LAMBDA;
```

Figure 9.12 Narrative input model for a triad with one spare.[23]

easier to define than the many statements in an equivalent input model, ASSIST offers a substantial enhancement to any SURE user.[24]

SURE computes the upper and lower bounds of system reliability. It does not compute an exact reliability, but the computed reliability bounds are usually within 5 percent of each other. SURE is available through COSMIC.

Examining the Ilities

Every specialty has its abbreviations and acronyms. Digital avionics goes one step further and tosses in ilities for good measure. Digital avionics are rife with ilities, and it is helpful to examine them early since they serve as valuable yardsticks by which to assess a design. The importance of each ility depends on the system design and its application.

Earlier in this chapter the three most important ilities, capability, reliability and maintainability, were reviewed. However, there are many other ilities that warrant mention to complete the discussion.

Accessibility, testability, and repairability are related to maintainability. Since all avionics will require maintenance or repair at some time, accessibility is a major design factor.

Systems that are reliable and maintainable will yield high availability. Military and civil fleet operators pay special attention to availability of aircraft since it is a major determinant of mission effectiveness. Aircraft that have to be repaired often or take too long to repair are not contributing to the mission since they are not available to fly.

Manufacturability is a major issue in civil avionics where the number of units to be manufactured can be large. If a design is complex to manufacture, the acquisition cost will be too high and, consequently, will adversely affect the LCC.

Affordability is another name for cost. As avionics grow in capability and importance, their cost escalates. Consequently, more attention is being paid to the avionics share of the cost. There are many trade-offs to be made between design, development, acquisition, and operating costs.

Closely related to affordability are interchangeability and compatibility since systems with those attributes generally are more affordable. Although not usually major design drivers, wherever it is possible, interchangeability and compatibility with previous or other contemporary systems should be included, provided it can be achieved without sacrificing other, more important features. Retrofittability—the capability of a new design of equipment to be successfully installed and operated in place of older, less capable equipment—can be placed in this category also, although there are occasions when equipment is designed especially for retrofit applications.

An avionics systems design should always factor in supportability. Support costs are a major element of operating costs. Where possible, the design should use parts and support equipment common to other systems so that support costs can be spread across more than one system. Sound system design principles dictate that no unusual or unique circuits or parts be used that are expensive, difficult to keep in stock, or in danger of becoming obsolete after the system has been in service for only a few years.

Digital avionics intrinsically have three important abilities, flexibility, adaptability, and programmability, which are derived from the software that drives the digital hardware. Digital avionics are as flexible, adaptable, and programmable as the imagination of the designer and the computer memory allow. Software ranks with hardware in importance in an avionics system design and in recent years has been one of the largest factors in determining the cost of a system. Software is discussed in other chapters.

References

1. FAR Part 25.1309: Equipment, Systems and Installation. Federal Aviation Administration.
2. FAA Advisory Circular 25.1309-1A System Design and Analysis, June 21, 1988.
3. NUREG 0492 Fault Tree Analysis. Nuclear Regulatory Commission, 1981.
4. *Digital Systems Validation Handbook*, vol. II, Federal Aviation Administration, DOT/FAA/CT-88-10, 1989.
5. Aerospace Recommended Practice 1834: Fault/Failure Analysis for Digital Systems and Equipment, SAE, 1986.
6. MIL-STD-1629 Procedures for Performing a Failure Mode and Effects and Criticality Analysis, 28 November 1984.

7. MIL-STD-882B: System Safety Program Requirements, 1 July 1987.
8. Hair, K. A., Leonard, J. B., "The Implications of Electromechanical Actuation on Aircraft Survivability," AIAA paper no. 84-2744-CP, 1984.
9. DO-160C: Environmental Conditions and Test Procedures for Airborne Equipment, RTCA, 1989.
10. FAA Advisory Circular 21-16C: Radio Technical Commission for Aeronautics Document No. DO-160C, February 14, 1990.
11. MIL-STD-810E Environmental Test Methods and Engineering Guidelines, 9 February 1990.
12. MIL-STD-704E: Aircraft Electrical Power Characteristics, 1 May 1991.
13. MIL-STD-461C: Electromagnetic Emission and Susceptibility Requirements for the Control of Electromagnetic Interference, 15 October 1988.
14. MIL-E-6051D: Electromagnetic Compatibility Requirements, Systems, 26 February 1988.
15. Clarke, C. A., and Larsen, W. A., "Aircraft Electromagnetic Compatibility," DOT/FAA/CT-86/40, 1987.
16. MIL-E-6051: Electromagnetic Compatibility Requirements, Systems, 26 February 1988.
17. MIL-STD-462: Electromagnetic Interference Characteristics, Measurement of, 15 October 1987.
18. MIL-STD-463A: Definitions and Systems of Units, Electromagnetic Interference and Electromagnetic Compatibility Technology, 19 September 1988.
19. FAA Advisory Circular 20-136: Protection of Aircraft Electrical/Electronic Systems Against the Indirect Effects of Lightning: March 5, 1990.
20. MIL-HDBK-253: Guidance for the Design and Test of Systems Protected Against the Effects of Electromagnetic Energy, 28 July 1978.
21. MIL-B-5087: Bonding, Electrical and Lightning Protection, 24 December 1984.
22. Bavuso, Salvatore J., and Peterson, P. L., *CARE III Model Overview and User's Guide* (first rev.), NASA TM-86404, 1985.
23. Butler, Ricky W., and White, Allen L., *SURE Reliability Analysis—Program and Mathematics*, NASA TP-2764, 1988.
24. Johnson, Sally C., *ASSIST User's Manual*, NASA TM-87735, 1986.

Bibliography

Henley, Ernest J., and Kumamoto, Hiromitsu, "Reliability Engineering and Risk Assessment," Englewood Cliffs, NJ, Prentice-Hall, 1981.

Lala, Jaynarayan H., and Smith, Basil, III, *Development and Evaluation of a Fault-Tolerant Multiprocessor (FTMP) Computer, Vol. III: FTMP Test and Evaluation*, NASA CR-166073, 1983.

Chapter

10

Software Design, Assessment, and Validation

Software is the heart of modern digital avionics; it turns a collection of silicon and copper into a system that gives the aircraft extraordinary capabilities. Software also gives the system flexibility far beyond that from analog equivalents. It describes algorithms, logical statements, data and control processes, and other abstract concepts. Because of this fundamental difference when compared to the tangibility of hardware, the design guidelines and procedures are different for software vis-à-vis hardware. This chapter opens with a discussion of a generic software development procedure that provides an understanding of the processes and caveats that are a part of software development. In the following section some of the principles of computer-aided software engineering (CASE) tools are reviewed.

The designing, coding, and testing of software requires unique procedures to prove beyond a reasonable doubt that the software is as reliable and comparable in quality as its host hardware. Separate, but similar, procedures have been established for the development and acceptance of software for military and civil systems. The third and fourth sections of this chapter examine very important documents. The third section discusses RTCA document DO-178 Software Considerations in Airborne Systems and Equipment Certification for application to civil aircraft. The fourth section discusses DOD-STD-2167 Defense Systems Software Development for application to military aircraft. The fifth section examines the requirements of DOD-STD-2168 Defense System Software Quality Program.

Modern military digital avionics systems are required to use Ada as a higher-order programming language for embedded computer systems. The sixth section of this chapter reviews ANSI/MIL/STD-1815

Ada® Programming Language. Ada is now being used in civil aircraft such as the B-747-400 and the Beech Starship. The use of Ada is expected to grow exponentially as software analysts and programmers become proficient in the language and efficient, validatible compilers become more widely available.

Standardization offers the opportunity to reduce cost and enhance interchangeability. To spur the realization of these benefits in military computers, MIL-STD-1750 Sixteen-Bit Computer Instruction Architecture was established. The final section of this chapter presents the major features of the standard so the designer is aware of the attributes of this popular instruction set.

The Essentials of Software Development

Software comprises algorithms, control and data flows, and other abstractions. Therefore, it cannot be as readily visualized as the hardware in a digital avionics design. This fundamental difference between hardware and software means that high-quality software is achieved only through a highly structured and controlled development process that includes extensive analysis, documentation, and testing.

This section describes a generic software development process and its elements to enable the reader to understand the process without being encumbered by esoteric nomenclature and nuances found in the highly specific procedures applied in developing software for civil and military systems. These specific civil and military software development procedures are presented in later sections of this chapter. The present section also includes some software design tips.

The flow of software development begins with the top-level system requirements and continues through the division between hardware and software functions. From there it continues to detailed software requirements, coding, testing, and finally, integration with the hardware to form the system. This process is overlaid with several layers of configuration control and reviews based on the criticality of the function performed by the software.

Successful software development begins with a clear understanding of the system performance specification and its subsequent accurate interpretation into software requirements. The designer must understand the system specification thoroughly to generate valid software requirements. Every requirement at each level of software must be traceable back to the system specification. Traceability is enhanced by using a numbering system in which each level of more detailed requirements adds a digit to the requirement identification number. Data flow and control flow diagrams are useful in visualizing the software design and in preparing accurate and understandable requirements. It is important to remember that requirements should state what needs to be done, not how to do it. Many of these points are explained in greater detail in Chaps. 1 and 3 and in the next section.

A top-down approach that is widely used in successful software design begins with the highest-level abstraction of the software and cascades downward into greater detail. It is important to complete the preliminary design to the module level (the smallest element of a program that can be uniquely identified and which performs a function or group of functions) as soon as possible for the early identification of any particularly difficult interfaces or modules. A top-down design enhances traceability of requirements and understanding of the software structure. As part of the top-down design, the trade-off between software and firmware is made, program memory resources are allocated, and interface and timing requirements are set. Common databases are also identified.

Sound design practice suggests that at this stage no more than 50 percent of the available memory should be used to ensure that there is room for the inevitable growth during development and after the system enters service. Because memory hardware costs are so low compared to making modifications to add memory after the equipment is in service, prudent design principles would suggest an even larger memory margin.

Interfaces and partitioning are complementary issues. A well-partitioned design with carefully chosen interfaces simplifies the modules, eases the process of making changes in the code, and improves the certificability in civil systems. A major objective of partitioning is to achieve a capability that permits changes to be made to a single module without requiring related changes to other modules that interface with it. Interfaces must be carefully documented, preferably in a tabular form such as a data dictionary that shows data flow from the originating module to the destination module(s).

Timing requirements include rates of receiving inputs and generating outputs and response times. These timing requirements translate directly into some often very demanding processor throughput requirements. For avionics functions that require very fast response, such as flight controls and displays, there can be no compromise on throughput requirements. Many customers will require a substantial margin in throughput capacity during the early design stages of a project. As a guideline, for military avionics systems a throughput margin of 50 percent is recommended at the time the Preliminary Design Review is held. Since processor costs are low relative to other costs in a typical program, prudent design principles would suggest an even larger margin.

Along with the design of the software, a test plan must be developed. Well-written test plans include exhaustive testing of each module since fault-free software begins with modules that can be shown, beyond a reasonable doubt, to be free of errors. At the module integration and higher levels of testing, the plan specifies the sequence of testing, initial conditions, and representative test cases. In a properly prepared plan, each test can be traced back to software and system requirements.

As a general rule, the software designers should not write the test plan for the software they develop except for testing at the module level. Above the module level, the test plan should be written by personnel who have no role in the design or coding of the software.

Another early task is the establishment of a software development plan. This plan must present all aspects of software management including configuration control, documentation, testing, schedule, and resources (both financial and personnel). Configuration control prescribes the level of control applied to the software at various stages of development as well as the procedures to be followed for correcting errors and implementing changes. The development plan must state precisely what documentation will be delivered, in what form, and on what schedule. Typical documentation includes not only the host machine code but test results and an operator's manual as well.

Upon completion of the software design requirements, test plan, and development plan, a Preliminary Design Review is usually held to permit the customer to examine and approve the documents. This review helps to ensure that the software developers have properly interpreted the system requirements and have established adequate procedures for developing high-quality software.

Coding marks the beginning of the software build. High-quality software begins with the modules. Well-designed modules are moderately sized (no more than a few hundred lines of code), efficient, and have well-written descriptive comments and few interfaces.

Where modules include functions that also appear in other modules, identical codes for that function should be replicated in all modules. Although the replication of codes substantially reduces development cost, it also introduces the risk of a latent error appearing in multiple locations in the software design. However, the risk of latent errors is reduced by using straightforward coding and thoroughly testing each module. Furthermore, replicated code means that an error may be detected sooner since the replicated code will be exercised more often than single-function code.

The next step in the software build is testing. The goal of module testing is to prove, beyond a reasonable doubt, that the code is error-free. Every function, mode, and logical state of the module must be tested. Testing must examine the performance of the module in the absence of valid data and check all algorithms in regions of discontinuities. Performance with data within, outside, and at the boundary of specifications must be checked. Run-time must be determined.

As modules are brought together in integration testing, the emphasis shifts to wringing out the interfaces, checking timing, and examining performance in the presence of a fault in an interfacing module.

Before integrating the software with the hardware, a test readiness is generally held to examine the module and integration test results and to give the go-ahead for the integration with the hardware. The emphasis at this review is on the thoroughness of testing to date, results, and the resolution of any anomalies identified as a result of the testing.

Testing the software after it is integrated with the hardware focuses on the hardware response to commands and the software performance in the presence of a hardware failure. Interrupt handling and bus and other resource sharing is examined and confirmed.

Some Helpful CASE Tools

The general software development discussion above suggests the need to bring order to what could be almost a random process. In recognition of this need, structured methods were developed and have evolved into CASE tools to bring discipline and rigor to the software design process and automate many of the repetitious development functions. The strength of CASE tools lies in their intrinsic checks and balances and consequent rigor. CASE tools are based on the fundamental precepts that data must be identified before it can be controlled and that data is the manifestation of system performance and the key to system control.

Structured methods, the foundation of CASE tools, breaks the software development process into two fundamental phases: structured analysis to develop the specifications for the software and structured design to produce the software architecture, and, in some of the more capable CASE tools, the software code.

An excellent introduction to structured methods can be found in *Advanced Integrated Flight Control System*.[1] Figure 10.1 summarizes the method described by Frenock. Structured methods begin with requirements, developed in earlier stages of the design process, stated in terms of functions, interfaces, and modes. Structured analysis uses these requirements to develop the environmental model that contains the data context and control context diagrams and continues with the development of data flow and control flow diagrams, control and process specifications, and the information dictionary that comprise the

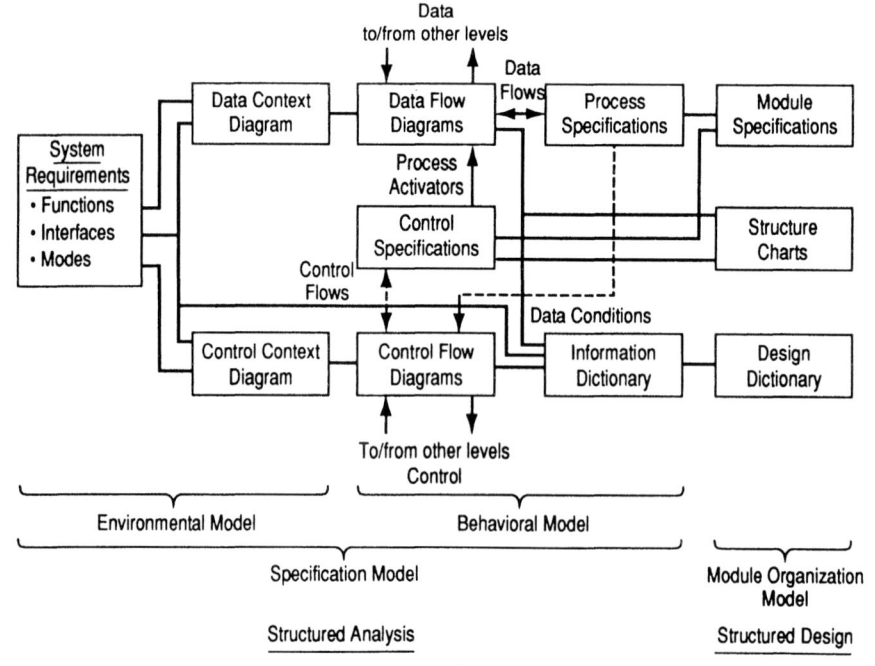

Figure 10.1 Structured methods flow diagram.[1]

behavioral model. The environmental and behavioral models combine to form the specification model. Structured design uses the specification model to build the module organization model that includes module specifications, software structure charts, and the design dictionary. Another excellent source for introductory material on structured methods is *Strategies for Real Time System Specification*.[2]

There is no single standard structured method; however, there is general (but not unanimous) agreement on the basic terminology and symbology as shown below:

Element	Symbol	Named by
Process	Circle	Verb and object
Data flow	Solid vector	Noun and adjective
Data store	Parallel lines	Noun and adjective
Control flow	Dashed vector	Noun and adjective
Terminators	Rectangle	Noun and adjective

From a structured methods perspective there are two types of signals in a system. If the signal is used in computation, it is data and is shown on the data flow diagram. If the signal is used to control the system, it will be shown on the control flow diagram. If the signal is both data and control, it will be shown on both diagrams and given a common name.

Figure 10.2 is a typical data context diagram that captures the essentials of a system from a data perspective. Note the use of the symbology described earlier. For a given system, there will be only one data context diagram.

Figure 10.3 is a typical top-level data flow diagram that gives a very general depiction of data flow in the system. In a process known as leveling, that data flow diagram is decomposed into lower levels of data flow diagrams, with multiple diagrams at each level, until the lowest possible level, which is a functional primitive. The functional primitive is documented in a process specification (PSPEC) that includes a brief narrative description, equations, and/or diagrams as appropriate and is usually less than a page.

Figure 10.4 is a typical control context diagram that captures the essentials of a system from a control perspective. Note the use of the symbology described earlier and the similarity to the data context diagram. For a given system, there will be only one control context diagram.

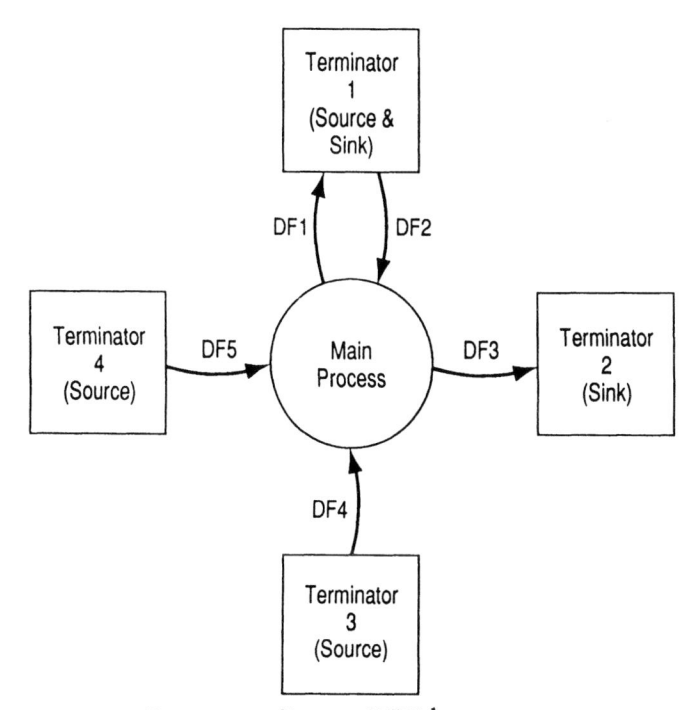

Figure 10.2 Data context diagram (DCD).[1]

198 Chapter Ten

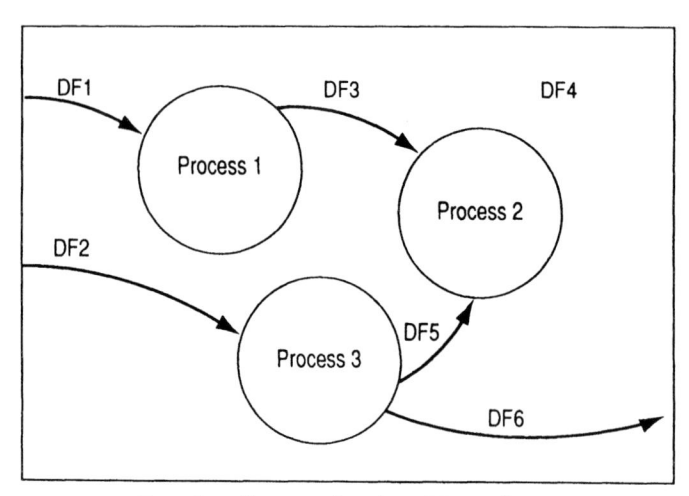

Figure 10.3 Data flow diagram, Level 1; (DFD 0).[1]

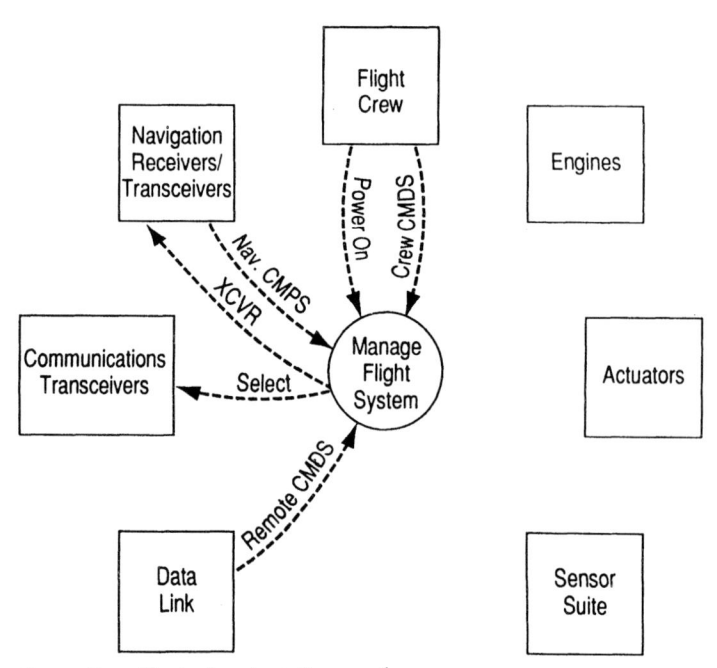

Figure 10.4 Control context diagram.[1]

Software Design, Assessment, and Validation   199

Figure 10.5 is a typical top-level control flow diagram that gives a very general depiction of control flow in the system. Note the similarity to the data flow diagram. The leveling procedure decomposes this control flow diagram into lower-level diagrams, with multiple diagrams at each level, until the lowest possible level, which is a control specification (CSPEC). CSPECs contain decision tables, state transition diagrams, and activation tables. A CSPEC is similar to a PSPEC and couples data flow diagrams (DFDs) with control flow diagrams (CFDs). Generally there is a CFD corresponding to each DFD, but if a DFD requires no control, the CFD can be omitted.

Structured design uses the output of the structured analysis phase to generate structure charts that show the hierarchical relationships

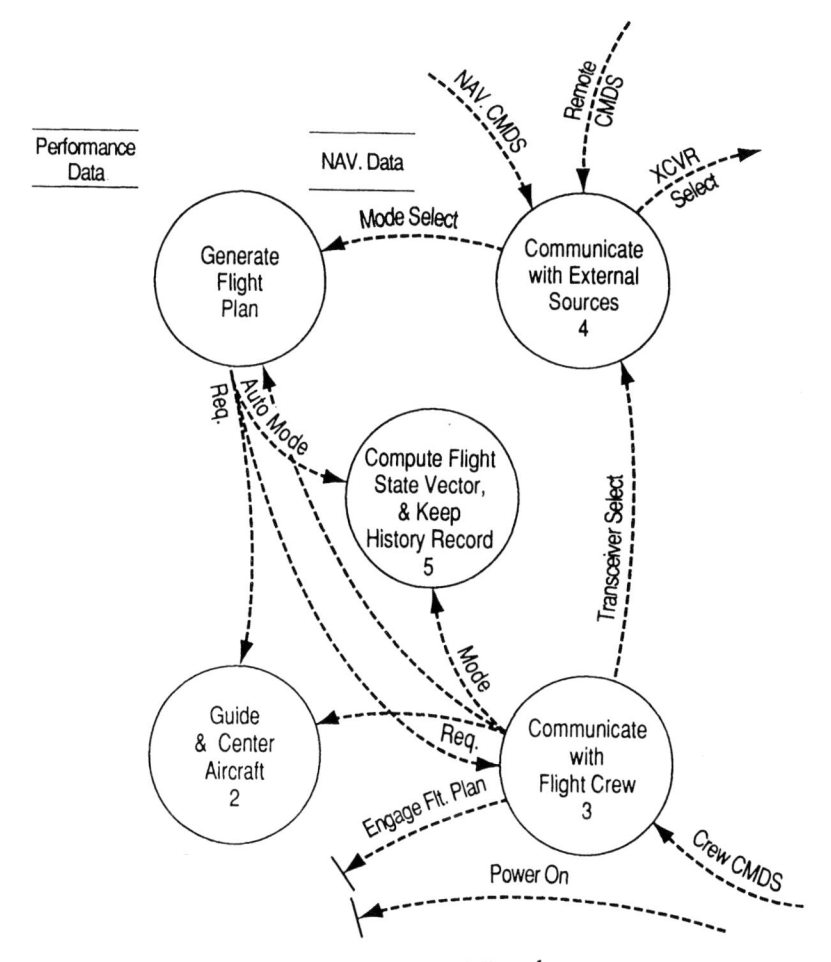

Figure 10.5   Control flow diagram, Level 1; (CFD 0).[1]

and interfaces for the software modules. The structure charts are developed from the DFDs. The data flow names and numbers match those on the DFDs. It is important to understand that structured design is an iterative process with structured analysis. As the design is developed, errors and omissions in the specification may be detected, and it will be necessary to change the specification by repeating some of the structured analysis procedure.

Modules are the key to a sound, functional software design. They must be understandable, have a single well-defined function, and be traceable to source DFDs and CFDs. They should be loosely coupled to permit independent development, testing, and changes to be made after the software has entered service. Upper-level modules should be traceable to CSPECs, while the lower-level modules should be traceable to PSPECs (Fig. 10.6).

The design dictionary alphabetically lists all data and control flows, update rates of the flows, and the physical attributes of data such as range, units, and precision.

There are CASE tools available for developing many software applications, both batch and real time. A number of surveys have been conducted to classify and summarize the capabilities of the tools. One of the most comprehensive of these surveys is *Software Methodology Catalog*.[3] This catalog, which lists numerous tools, gives a complete description of them and includes details about their applicability, ease of use, hardware and software required, cost, and whom to contact for additional information.

Of the countless CASE tools available, the ones of most interest to avionics software designers come from one of four basic methodologies: Extended System Modeling Language (ESML), Harel, Hatley-Pirbhai, and Ward-Mellor. Earlier in this section the Hatley-Pirbhai methodology was presented as an introduction to structured methods. The Carnegie Mellon University Software Engineering Institute conducted a survey of its affiliates to formulate a comprehensive evaluation of these methods. Based on the results of the survey, all other factors being equal, there was a slight preference for the Hatley-Pirbhai method. This preference was driven by the "quality of the available text and the overall simplicity and flexibility of their notations and techniques."[4]

What DO-178 Requires

RTCA document DO-178 Software Considerations in Airborne Systems and Equipment Certification is the authoritative source for procedures to certify the civil avionics software. Like most RTCA documents, DO-178 achieves its preeminent role by being invoked by the

```
           MODULE SPECIFICATION for LIMIT COMMANDS

INPUTS:   Data signal MODE INFO from DEVELOP
          FEEDFORWARD SIGNALS module.

          Data signal AUTOMODE INFO from DEVELOP
          FEEDFORWARD SIGNALS module.

          Data signal CMDS LIMIT INFO from external.

          Data signal TRAJECTORY CMDS from external.

          Data signal MANUAL CMDS from external.

OUTPUTS:  Data signal LIMITED CMDS to DEVELOP
          FEEDFORWARD SIGNALS module.

          Data signal LIMIT PROCESSOR FAULT INFO to
          DEVELOP FEEDFORWARD SIGNALS module.

TRACEABILITY REFERENCE: Process Spec. 1.

MODULE DESCRIPTION:

A = MAGNITUDE (CMD LIMIT INFO)

IF AUTOMODE = TRUE AND MODE = FALSE
      THEN CMD INPUT = TRAJECTORY CMDS
           IF CMD INPUT ≤ -A
                THEN LIMITED CMDS = -A
                ELSE IF CMD INPUT ≥ A
                          THEN LIMITED CMDS = A
                          ELSE LIMITED CMDS = CMD INPUT
                     ENDIF
           ENDIF
      ELSE CMD INPUT = MANUAL CMDS
           IF CMD INPUT ≤ -A
                THEN LIMITED CMDS = -A
                ELSE IF CMD INPUT ≥ A
                          THEN LIMITED CMDS = A
                          ELSE LIMITED CMDS = CMD INPUT
                     ENDIF
           ENDIF
ENDIF
```

Figure 10.6   Example software module.[1]

regulatory agencies for certifying software in civil digital avionics systems. Petitioners to the regulatory agencies are not required to use the software certification procedures in DO-178. However, the petitioners' proposed alternative procedures must be shown to be equivalent or better.[5]

The central theme in DO-178 is a disciplined approach to software definition, development, testing, and configuration management to yield software that is traceable, testable, and maintainable.

DO-178 follows the precedent established in the Federal Aviation Administration (FAA) Advisory Circular 25.1309-1A that recognizes three levels of criticality for aircraft systems: critical, essential, and nonessential (see Chap. 1). Just as 1309 calls for critical, essential, and nonessential hardware, DO-178 calls for equivalent levels of software: Level 1 is critical software, Level 2 is essential, and Level 3 is nonessential. Where the criticality of a function performed by a software module changes as a function of mission phase or aircraft state, the module must be developed and validated at its highest level of criticality.

DO-178 divides the generation and delivery of software into three phases: development, verification, and assurance. A flow chart depicting the development and verification phases is shown in Fig. 10.7. The top row shows development activities, and the bottom row shows the related verification activities. The related documents or codes are in the center row. It is important to note that every development activity has a matching verification activity.

The flow in Fig. 10.7 is from system design requirements to a validated total system code. There are several major points for the designer and programmer to keep in mind while following this flow. To begin with, the system design requirements must be verified to be correct since any error may generate a potentially catastrophic error in the software. Given a correct set of system requirements, software requirements must be developed and verified. In developing the software requirements, the same general guidance in developing system requirements applies—they

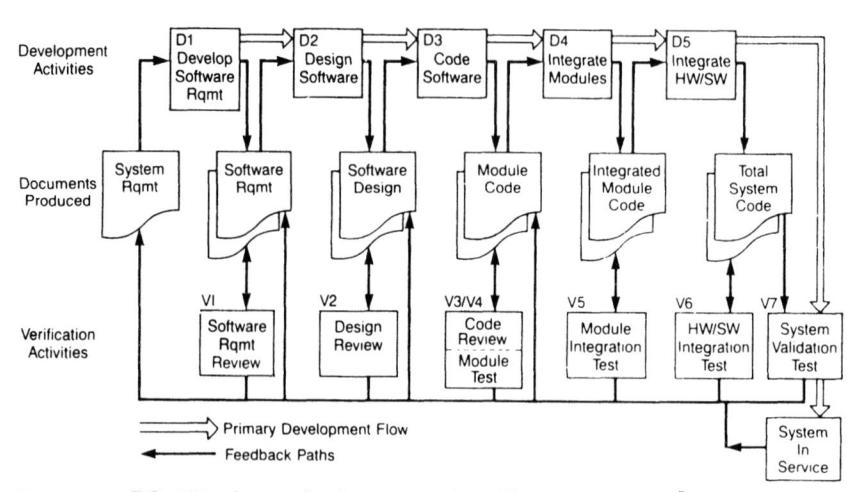

Figure 10.7 DO-178 software development and verification activities.[5]

should state what needs to be done, not how to do it. Typical requirements included are the function to be performed; criticality of the function; timing; hardware and software interfaces; partitioning; performance under fault conditions; and testing.

The software requirements serve as the basis for the software design. To repeat an earlier point, it is most important that the design be understandable, traceable to the design requirements, maintainable, and testable.

With the software design verified, coding of the software modules can begin. Verification of the coding yields an error-free compilation that is consistent with the applicable standards and design documents.

Throughout the first three stages of software requirements definition, design, and coding there are identical assurance requirements. For Level 1 software the results of the verification process, including all problems identified and their resolution, must be noted and retained. Traceability matrices that relate a requirement or attribute to a higher-level requirement must be developed. Task completion should be audited. For Level 2, a summary of the verification process is required along with a Statement of Compliance. These items may be a part of the Accomplishment Summary document, discussed later. Like Level 1 software, traceability matrices are required and task completion should be audited. Level 3 software requires no assurance measures.

Software test requirements are derived from the Software Requirements Document and the structure of the software. Testing begins at the module level and progresses to the integrated module and hardware and software integration testing. Where possible and feasible, testing should be done on the target machine to avoid repetition of selected tests when the software is installed on the target machine. At the module level the tests focus on logic and computation performance. At the integrated module level, tests focus on module linkages such as data and control flow, timing, sequencing, and partitioning.

Hardware and software integration testing emphasizes merging the software into the target hardware and verifying performance in this environment. Testing frequently takes the form of real-time simulations where carefully selected (on the basis of failure modes and effects analysis, or FMEA, and other analyses) faults, failures, and transients are imposed to determine the performance in off-nominal conditions. Integration testing is especially valuable in ferreting out errors such as incorrect handling of faults and transients, timing incompatibilities, and bus contention problems.

In software testing, the assurance requirements differ from those in earlier development phases. Levels 1 and 2 software require a requirements coverage analysis to identify the test case(s) which demonstrate(s) compliance with each software requirement and possible interactions among the requirements. Additionally, Level 1 software requires a software structural coverage analysis that identifies the test case(s) that executed each element of the software. Levels 1 and 2 software assurance requires that all test plans, procedures, and results, as well as the identification and resolution of all problems encountered in the testing, be recorded and retained. Level 1 also requires an audit of all tasks as done in the earlier phases of development. There are no assurance requirements for Level 3 software.

Demonstrating that the system complies with the top-level requirements is the final step in the software development process. By the time these tests are conducted, the software should be relatively free of errors. However, at this point errors that do occur can require substantial effort to correct.

Documentation is a central part of every software development program. Special attention is devoted to the documentation so that the software can be traceable, comprehendible, and maintainable. DO-178 calls for up to 14 documents in a complete software documentation package. A description of each document is provided here in Part A of App. B. Six of these documents are of particular importance: System Requirements, Software Requirements Document, Software Configuration Management Plan, Software Quality Assurance Plan, Configuration Index Document, and the Accomplishment Summary.

As the name suggests, the Systems Requirements Document is the top-level document from which all other requirements are derived. It includes a description of the system broken down by line replaceable unit (LRU), certification requirements, and needs for specific design techniques such as partitioning or redundancy.

The Software Requirements Document is derived from the systems requirements. The functional and operational requirements for the software are stated in quantitative terms along with graphical depictions of the relationships among the software modules. Criticality, timing, interfaces, and performance under fault conditions are included.

The first section of this chapter discussed software configuration management plans and their importance to a sound software development program. A Software Configuration Management Plan is required by DO-178, and its contents match the contents suggested in the earlier section. The topics included in the plan are configuration identification and control, configuration status accounting, configuration reviews and audits, and supplier control.

The Software Quality Assurance Plan presents the policies and procedures necessary to ensure that the software meets the quality standards imposed on its development. The plan contains details on quality assurance reviews and audits, problem reporting and corrective action methods, media control, and record keeping.

The Configuration Index Document (CID) is the major control document for each software version and provides the index of all applicable documentation. CIDs exist at both the system and LRU levels. The system CID lists the LRU names, hardware and software part numbers, and the LRU CIDs. The LRU CID contains the exact configuration of the software in the LRU and the applicable documents used in its development.

In dealing with the certification authorities, the single most important document is the Accomplishment Summary. It is not more than 10 pages long and contains as a minimum (1) a list of all other documents that may be submitted or information relevant to software certification, (2) a description of the system hardware and software, (3) criticality categories, and (4) the software verification plan and results.

What DOD-STD-2167 Requires

DOD-STD-2167 establishes "requirements to be applied during the acquisition, development, or support of software systems" for military use.[6] DOD-STD-2167 tightly couples hardware and software development, as shown in Fig. 10.8. This is the well-known waterfall chart of software development. Notice the separate branches of hardware and software development after the System Design Review (SDR) until the branches merge in system integration and testing.

The standard recommends a top-down software design approach and presents a hierarchy of software system elements as shown in Fig. 10.9.[6]

System/Segment
   Computer Software Configuration Item (CSCI)
      Computer Software Component (CSC)
         Computer Software Unit (CSU)

A system/segment specification (SSS) is a system-level requirements specification. A CSCI is a configuration item for computer software. (See App. A for the DOD-STD-480 definition of a configuration item.) A CSC is "a distinct part of a CSCI and may be further decomposed into other CSCs and CSUs." A CSU is "an element specified in the design of a CSC that is separately testable."[6]

Figure 10.10 recasts Fig. 10.8 to highlight the documents required as part of the software development process. The top bar contains,

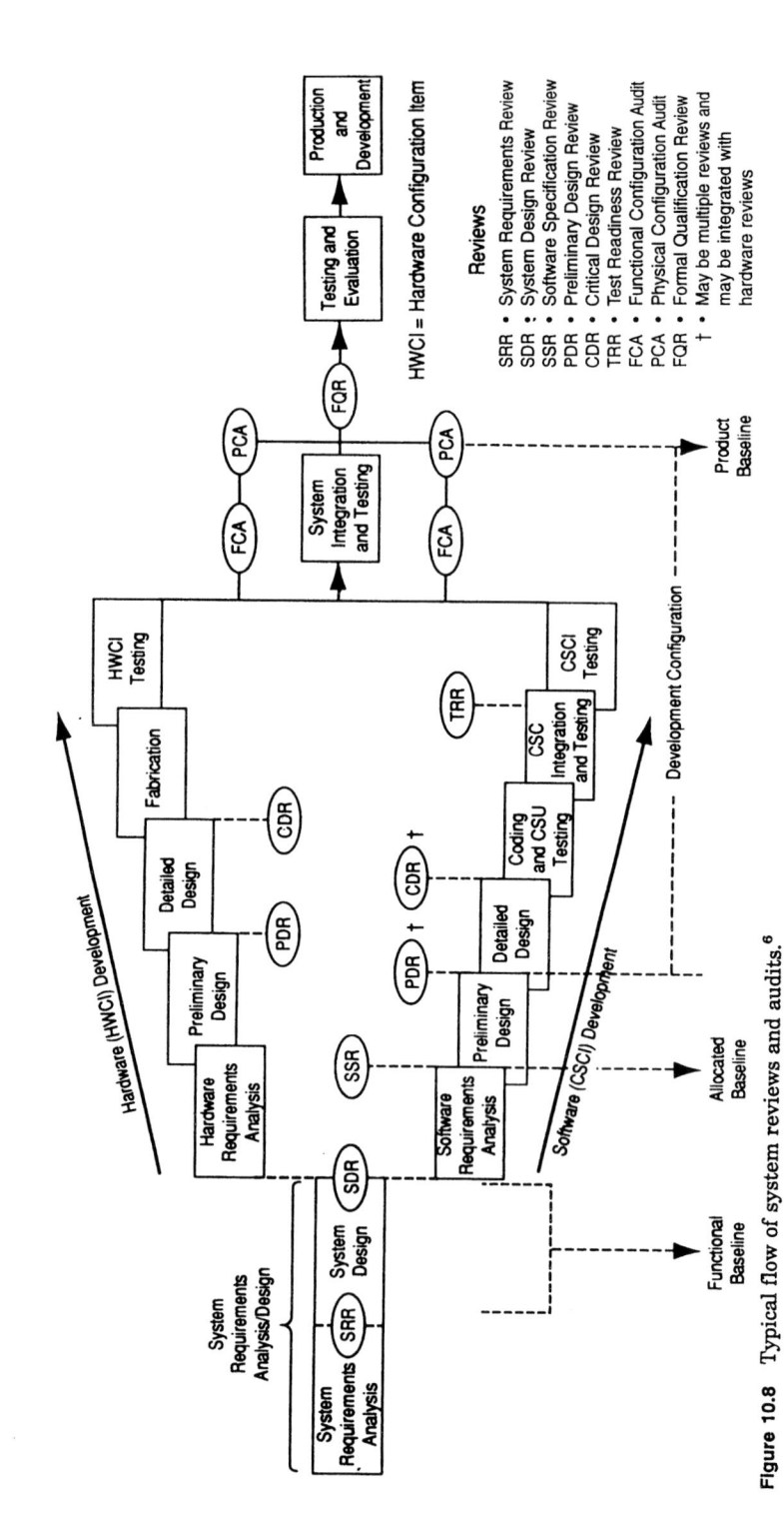

Figure 10.8 Typical flow of system reviews and audits.[6]

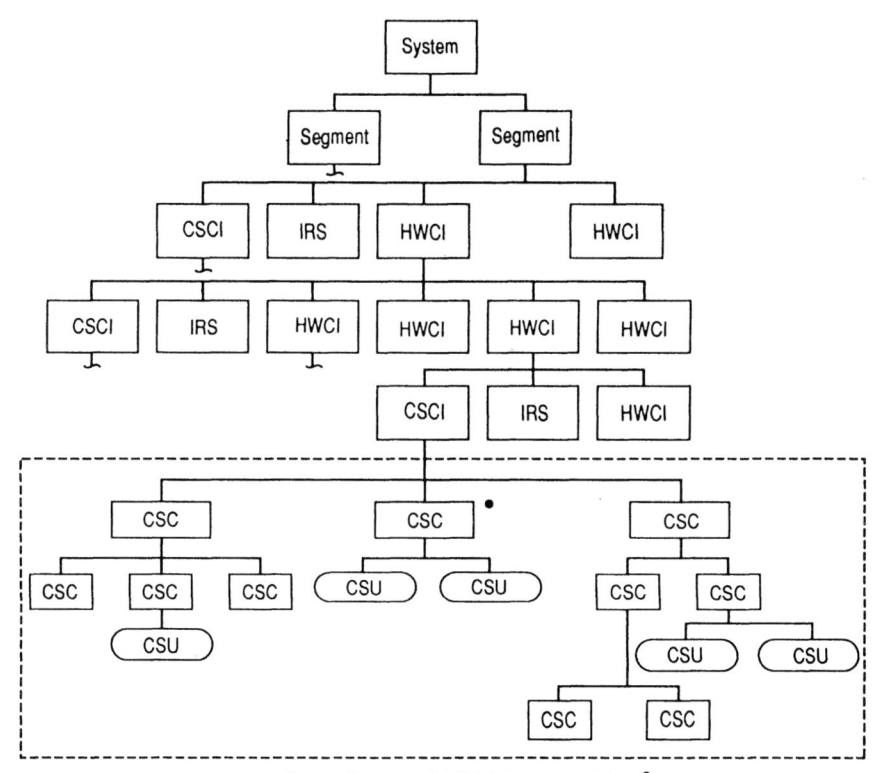

Figure 10.9 Typical system breakdown and CSCI decomposition.[6]

from Fig. 10.8, the various phases of a software development project and the associated reviews, and audits at the end of a phase are shown near the bottom. 2167 can require up to 18 documents to be prepared; however, tailoring the requirements of the standard may reduce that number. Notice that in many cases a document is required in both preliminary and final form (e.g., the system specification, software requirements specification, and so forth). The Functional Baseline is established upon completion of the System Design Review. The Allocated Baseline is based on the final Software Requirements Specification and the Interface Requirements Specification. The Product Baseline reflects the software configuration after the functional and physical configuration and prior to beginning system integration and testing. A detailed description of each document can be found in App. B.

2167 presents a number of evaluation criteria to be applied to the many products it requires. The documents must be internally consistent, understandable, and traceable to and consistent with higher-level documents. Development of the software must be in accordance with the analysis, design, and coding techniques stated in the Soft-

208 Chapter Ten

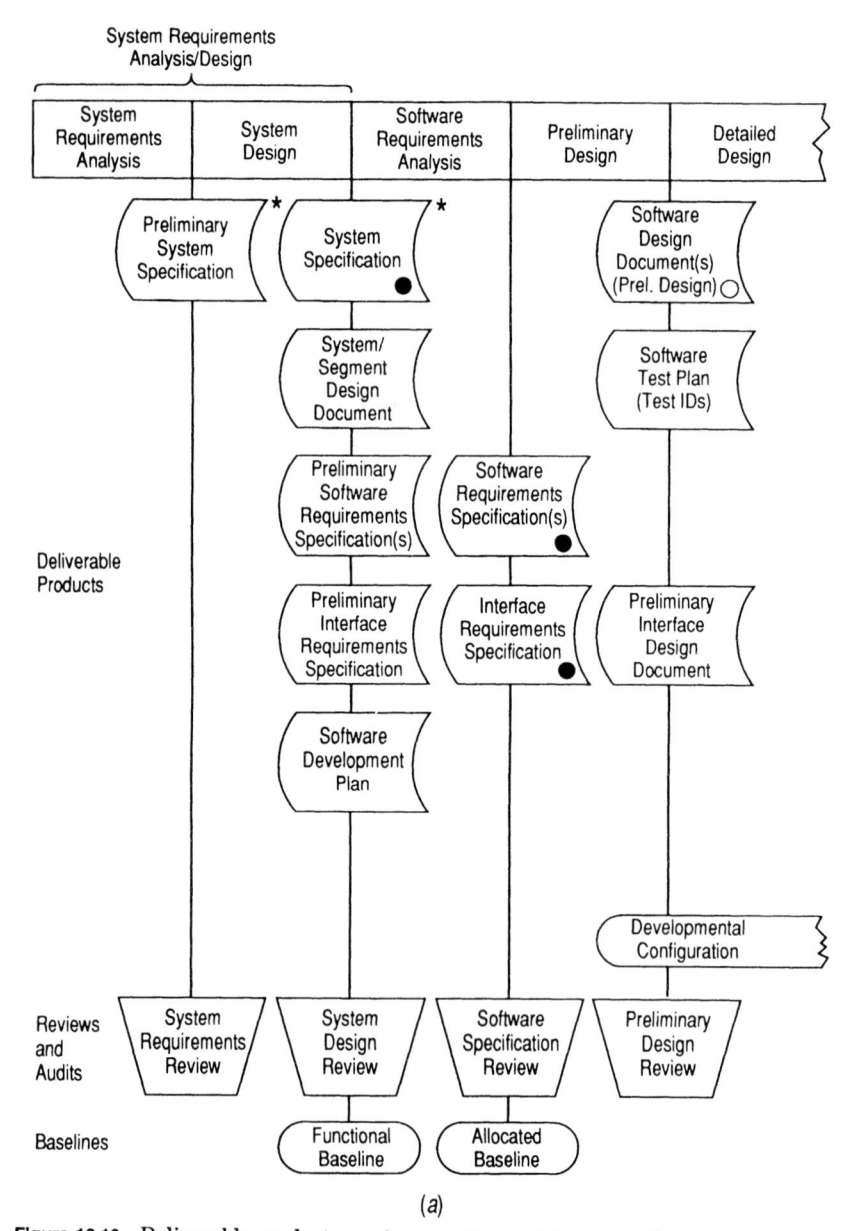

Figure 10.10 Deliverable products, reviews, audits, and baselines.[6]

Software Design, Assessment, and Validation 209

Detailed Design	Coding and CSU Testing	CSC Integration and Testing	CSCI Testing	System Integration and Testing
Software Design Document(s) (Det. Design) ○	Source Code Listings ○			
	Source Code		Updated Source Code	
Software Test Description(s) (Cases)		Software Test Description(s) (Procedures)	Software Test Report(s)	
Interface Design Document			Operation and Support Documents ★	

Notes:
● Incorporate into Baseline
○ Incorporate into Developmental Configuration
★ May be Vendor Supplied
* May be a:
  1. System/Segment Specification
  2. Prime Item Specification
  3. Critical Item Specification
† May be Deferred Until After System Integration & Testing

Developmental Configuration

| Critical Design Review | | Test Readiness Review | CSCI Functional & Physical Configuration Audits † | Version Description Document(s) † |

Software Product Specification(s) ● †

Product Baseline

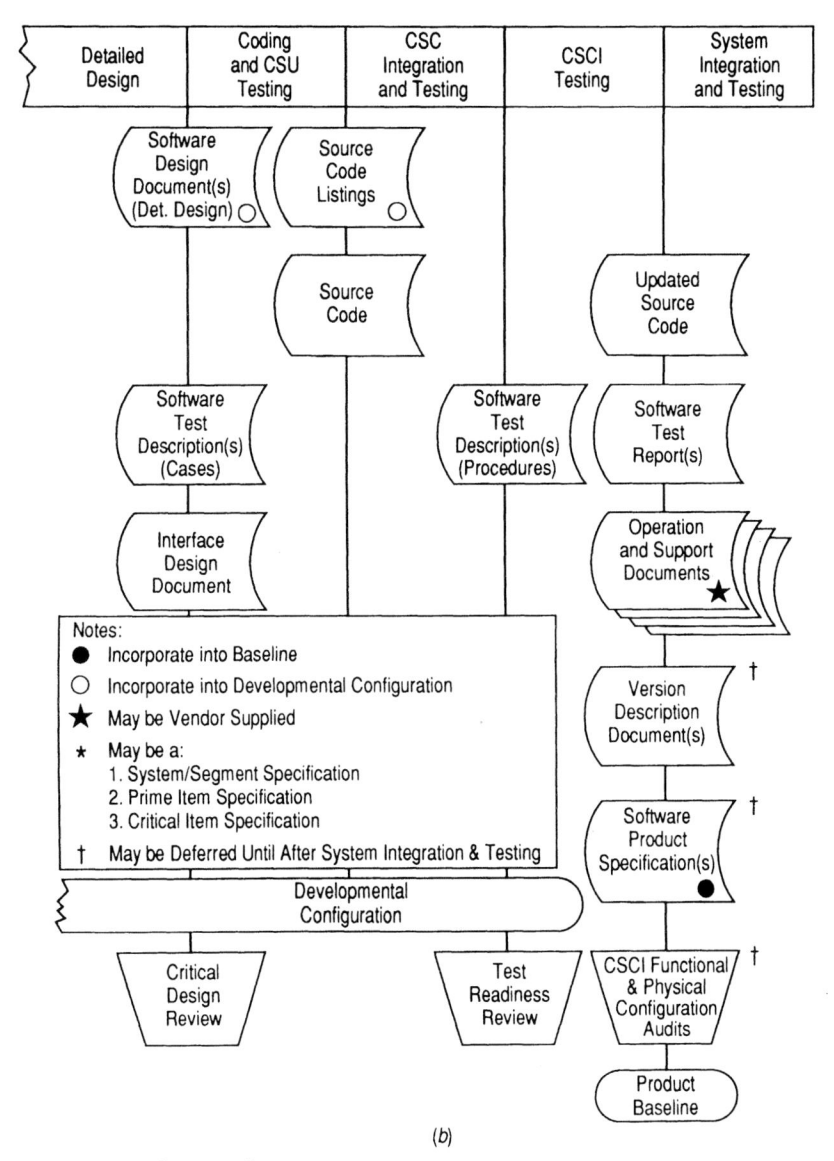

(b)

Figure 10.10 (*Continued*)

ware Development Plan. The size of a given software element and its execution time must not exceed the values established in the allocated baseline.

The contractor must prepare coding standards for review and approval by the customer which address:

- *Presentation style:* Indentation and spacing, capitalization, uniform presentation of information, use of headers, layout of source code listings, conditions and format for comments, and size of code aggregates (typically 100 to 200 executable, nonexpandable statements per CSU)

- *Naming:* Rules and conventions for selecting identifiers; restrictions on reserve and keywords

- *Restriction on the implementation language:* Restrictions imposed on using constructs and features of the implementation language due to project or machine-dependent characteristics

- *Use of language constructs and features:* Allowed usage of implementation language constructs and features

- *Complexity:* Controls and restrictions on the complexity of code aggregates.

In order to ensure that problems encountered in developing the software receive the proper level of attention, a taxonomy has been established. Problems are categorized according to the following descriptions:

- *Software:* Software does not operate according to the documentation, which is correct.

- *Documentation:* Software operates correctly, but the documentation does not describe the correct operation.

- *Design:* Software operates according to the documentation, but a performance deficiency exists.

Once the problem is categorized, it is assigned a priority according to the following guidelines:

- *Priority 1:* Prevents the accomplishment of a mission or operation-essential requirement or jeopardizes personnel safety

- *Priority 2:* Adversely affects the accomplishment of a mission or operation-essential requirement and no alternative, work-around solution is known

- *Priority 3:* Same as Priority 2 except an alternative, work-around solution is known
- *Priority 4:* Does not affect mission or operational requirements; an inconvenience or annoyance
- *Priority 5:* All other errors.

What DOD-STD-2168 Requires

The DOD-STD-2168 Defense System Software Quality Program implements a defense software quality program as required by Department of Defense Directive 4155.1. Software quality is defined as the ability of a software product to satisfy its specified requirements. Software quality programs must be established in accordance with 2168 "to assure the quality of (a) deliverable software and its documentation, (b) the processes used to produce deliverable software, and (c) non-deliverable software."[7]

The contractor must develop a Software Quality Program Plan (SQPP) as described in Data Item Description DI-QCIC-80572, which is part of the standard. The SQPP must discuss the software quality organization, personnel, schedule, procedures and tools to be used in developing and testing the software.[7]

Specific topics to be covered in the SQPP include the contractor's self-monitoring and evaluation of software, documentation, development processes (engineering, qualification, configuration management, and so on), software development library, nondevelopmental software, nondeliverable software, deliverable elements of the software engineering and test environments, subcontractor management, and acceptance inspection and preparation for delivery.

2168 explicitly states, "The persons conducting the evaluation of a product or activity shall not be the persons who developed the product, performed the activity, or are the persons responsible for the product or activity."[7]

Using Ada

When it comes to higher-order languages for programming digital avionics systems, there is only one choice: Ada.* There are, of course, other languages, each with its own advantages. However, for all Depart-

---
*ANSI/MIL-STD-1815 requires the first mention of Ada in any published material be accompanied by a footnote as follows: "Ada is registered trademark of the U.S. Government (Ada Joint Program Office)."

ment of Defense avionics systems designed and built today, Ada is the required language. The use of Ada was mandated by DOD Directive 5000.29: "The Ada programming language shall become the single, common, computer programming language for Defense mission-critical applications."[8] This mandate was reaffirmed in DOD Directive 3405.2, which also invoked DOD-STD-2167 for software development.[9]

Because of its proven performance during increasing use in military systems, Ada is also being applied in civil systems. In fact, airlines have "recommended that the Ada programming language be used as the standard high-order language (HOL) in avionics equipment design."[10] ARINC Report 613 prescribes the use of Ada in all avionics built in accordance with ARINC characteristics. Ada also has been mandated for the space station *Freedom*.[11] Thus, it is clear that digital avionics systems designers must be familiar with Ada so its features can be used to the best advantage.

The growing use of Ada comes from both the stick and the carrot approaches. The stick has just been described in the form of DOD directives and ARINC reports. Equally important, though, are the carrots in terms of anticipated benefits from using Ada. These benefits include code portability and reusability, increased programmer productivity, ease of reassigning software personnel from project to project, increased reliability (fewer design and coding errors) and maintainability, and improved management visibility into the software development process.[12]

The basic reference for Ada is ANSI/MIL-STD-1815: Ada® Programming Language. As shown by the title, this military standard also has been adopted by the American National Standards Institute (ANSI). It has been adopted by the International Standards Organization (ISO) as well.[13]

Ada is an extremely powerful and versatile language. Some of its most notable features include floating point arithmetic and the capability to permit the user to see the types of information available from a program while hiding the implementation details. Like most higher-order languages, Ada reduces programming costs because its English-like constructs expedite both the writing and checking of the code.

The basic element of an Ada program is a program unit which consists of subprograms, packages, or tasks. Execution of the program is accomplished by executing a subprogram called the main program, which may invoke other subprograms declared in other compilations units.

Subprograms contain algorithms that describe either procedures or functions. A procedure invokes a series of actions such as reading data, updating variables, or generating an output. A function is very similar to a procedure except that a function also returns a result.

A package specifies a group of logically related entities, such as types, objects of those types, and subprograms with parameters of those types. It is written as a package declaration and a package body. The package declaration has a visible part containing the declarations of all entities that can be explicitly used outside the package. It may also have a private part containing structural details that complete the specification of visible entities but which are irrelevant to the user of the package. The package body contains implementations of subprograms (and possibly tasks as other packages) that have been specified in the package declaration.

A task operates in parallel with other parts of the program. It is written as a task specification (which specifies the name of the task and the names of formal parameters of its entries) and a task body which defines its execution.

Ada has been proven in a mission critical flight control system application with several pathfinding results. Figure 10.11 shows the programmer productivity gains from using Ada when compared to assembly languages in the design, coding, and testing of control laws. As the use of Ada expands and programmers become more proficient in its use, Ada should require only one-fourth as many programmer hours as floating point assembly language to implement comparable functions.[14]

Figure 10.12 offers some insight into the reason for the gains in programmer productivity when using Ada. The figure shows the trans-

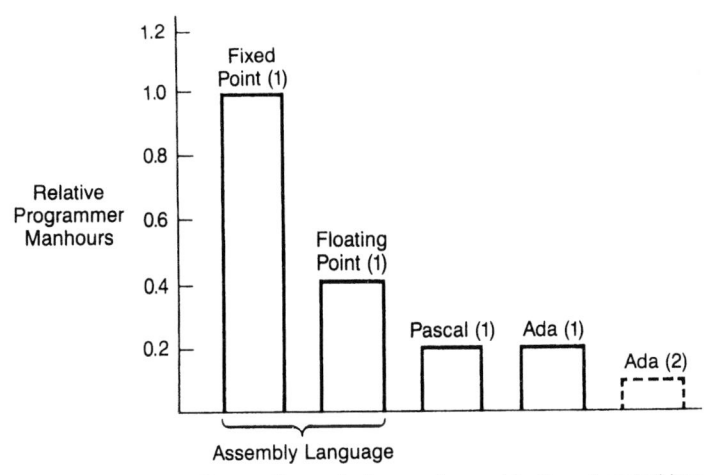

(1) Effort applies to design, coding and testing of control laws
(2) Pervasive use of Ada estimate

Figure 10.11 Impact of higher-order language and floating point on programmer productivity.[14]

Original S-Plane Representation of Structural Filter

$$\frac{0.48075^2 + 83.5535S + 3894}{S^2 + 125S + 3894}$$

Assembly Language Statements*

STFL = [0.56503 * PRESTRU] - [0.33991 * PREM1STRU] + [0.089533 * PREM2STRU] + [0.87711 * STFLM1] - [0.19182 * STFLM2]

LDL	RR10, PCAS24;	
CALL	FMUI;	% RR8 Contains PRESTRU
LDL	RR6, RR8;	% 0.56503 * PRESTRU
LDL	RR8, PREM1STRU;	% Store Result for Later Use
LDL	RR10, PCAS25;	
CALL	FMUI;	% 0.33991 * PREM1STRU
LDL	RR10, RR8	% Prepare for Subtraction
LDL	RR8, RR6;	
CALL	FSUB,	% RR6 - [0.33991 * PREM1STRU]
LDL	RR6, RR8	% Store Result for Later Use
LDL	RR8, PREM2STRU,	
LDL	RR10, PCAS 26;	
CALL	FMUI;	% 0.089533 * PREM2STRU
LDL	RR10, RR6,	
CALL	FADD	% RR6 + [0.089533 * PREM2STRU]
LDL	RR6, RR8;	% Store Result for Later Use
LDL	RR8, STFM1;	
LDL	RR10, PCAS27;	
CALL	FMUI;	% 0.87711 * STFLM1
LD	RR10, RR6;	
CALL	FADD;	% RR6 + [0.87711 * STFLM1]
LDL	RR6, RR8	% Store Result for Later Use
LDL	RR8, STFLM2;	
LDL	RR10, PCAS28;	
CALL	FMUI;	% 0.19182 * STFLM2
LDL	RR10, RR8;	% Prepare for Subtraction
LDL	RR8, RR6;	
CALL	FSUB;	% RR6 - [0.19182 * STFLM2]
LDL	STFI, RR8;	

* Based on floating point, twice as long if fixed point

Difference Equation Representation

STFL = 0.56503 * PRESTRU
 0.33991 * PREM1STRU
 + 0.089533 * PREM2STRU
 + 0.87711 * STFLM1
 0.19182 * STFLM2

OR

Ada Statement

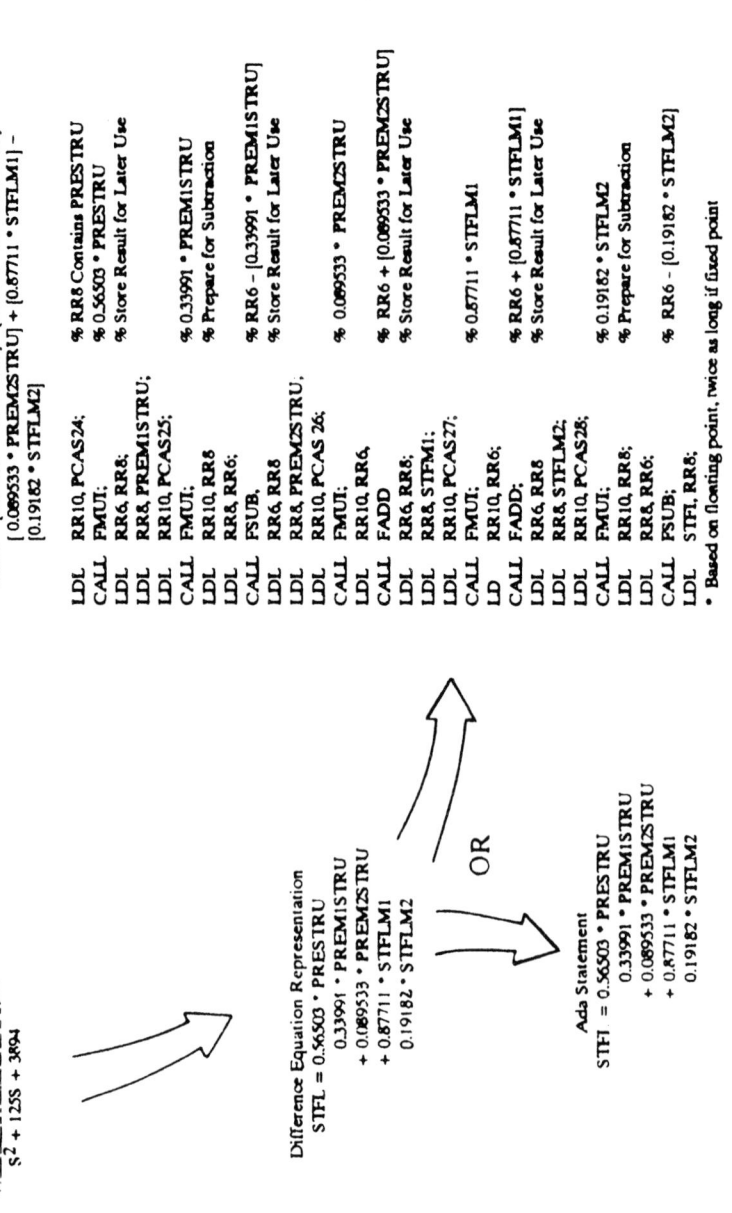

STF1 = 0.56503 * PRESTRU
 0.33991 * PREM1STRU
 + 0.089533 * PREM2STRU
 + 0.87711 * STFLM1
 0.19182 * STFLM2

Figure 10.12 Steps in digital mechanization of structural filter.[14] (© *American Institute of Aeronautics and Astronautics;* reprinted with permission.)

formation of a typical S-plane representation of a structural filter into a difference equation representation. This equation is then shown in equivalent forms of machine language and Ada statements. The easily readable and brief Ada statements are the direct result of Ada being a higher-order language and its ability to handle floating point operations.[14]

Table 10.1 shows the impact of Ada on different phases of software development. Under the operational flight program functional divisions column the percentages refer to the fraction of words devoted to that function in the flight program. The percentages under each phase of software development are the fraction of software development activities in that phase. Note that the impact of Ada is primarily a reduction in the coding phase for all functions. For the control law functions Ada also has a large impact on testing and a modest impact on design, which supports the point made earlier in Fig. 10.11. In the study which provided the data for this figure, Ada had virtually no impact in software design and maintenance.[14]

A major concern in using higher-order languages is overhead, or efficiency. Any gains in ease of programming made possible by Ada may be more than offset by requirements for additional processing time and memory. An early experiment with Ada on the F-15 Eagle showed a 10 percent growth in computation time and a 63 percent growth in memory capacity requirements when compared to flight control software written in machine language.[14] Thus, even though the use of Ada is mandatory in military avionics, the designer must be aware of the potential penalties associated with its use, especially for flight control systems where real-time performance is extremely important.

The Boeing Commercial Airplane Co. has conducted a thorough study of Ada and its potential application to large commercial passenger transport aircraft. Their conclusion is that Ada compilers have matured to the point where Ada can be used in a transport aircraft embedded computer environment without a performance penalty. As observed in the F-15 Eagle case, there is still a memory overhead pen-

TABLE 10.1 Impact of Ada on Software Productivity*[14]

Operational flight program functional divisions		Software development process			
		Design 25%	Coding 20%	Test 45%	Maintenance 10%
Control laws	20%	Some	Large	Large	Some
Executive	15%	Nil	Large	Some	Nil
Redundancy management	20%	Nil	Large	Some	Nil
Built-in test	45%	Nil	Large	Some	Some

*© American Institute of Aeronautics and Astronautics; reprinted with permission.

alty, but it appears probable that it can be reduced to an acceptable 20 percent. Ada should reduce life cycle costs also, particularly for programs of over 1000 delivered lines of source code.[15]

Ada has been successfully used in civil applications, the most notable being the Beechcraft Starship. The Rockwell International-Collins Co. built the electronic flight displays (EFDs) and the flight management system (FMS) for the Starship, and all of the software is written in Ada. The EFD has 31,000 lines of source code occupying 318,000 bytes of target memory, and the FMS has 51,000 lines of source code requiring 470,000 bytes of target memory. The software runs on Intel 80186 processors. It required 30 programmer years to develop and deliver the software, which translates into over 200 delivered source lines of code per month.[16]

There are, of course, some important issues that must be addressed if Ada is to be successfully applied. Foremost among these issues is that of compilers. All Ada compilers are not created equal as is clearly demonstrated in the paper by Bhansali, et al.[15] Many compilers generated code that was as efficient as Pascal, even when including the Ada exception handling capability. However, among Ada compilers there were substantial differences in code efficiency and in memory size requirements.[15]

All Ada compilers must be validated using the Ada Compiler Validation Capability (ACVC), a suite of over 4800 test programs (Version 1.11). When a new ACVC version is issued, all compilers must be revalidated. Compilers must also be revalidated if there have been substantial changes.[17]

Another major concern about Ada in flight critical applications is dynamic storage management. "The amount of computer memory used by an Ada program at any specific point in its execution can vary significantly depending upon the branches or code paths taken and the particular Ada statements executed to arrive at that point."[18] However, a solution that calls for a graded approach to using Ada has been developed to resolve this concern.[19]

One final concern about Ada is its nondeterministic behavior when there are multiple tasks executing on the same processor. Care must be taken to select a compiler which makes a deterministic but fair choice among entries, based on a repeatable algorithm.[18]

MIL-STD-1750 Instruction Set Architecture

In order to reduce the cost and enhance the interchangeability of computers used in avionics systems, the United States Air Force has directed the use of a standardized instruction set as established in MIL-STD-1750 Sixteen-Bit Computer Instruction Set Architecture. MIL-

STD-1750 establishes requirements for data and instruction formats and for addressing modes, registers, memory, interrupts, and input-output.[20]

An Instruction Set Architecture (ISA) is defined in the standard as:[20]

> The specification of the interface between the software and the hardware. It describes the conceptual structure and functional behavior of a computer as distinct from the organization of the data flow and controls, logic design or physical implementation. ISA includes the processor and input/output instruction sets, their formats, operations code, and addressing modes; memory management and partitioning if accessible to the machine language programmer; the speed of accessible clocks; interrupt structure; and the manner of use and format of all registers and memory locations that may be directly manipulated or tested by a machine language program.

There are four classes of requirements in 1750: mandatory, optional, spare, and reserved. Mandatory requirements are self-explanatory. Optional requirements apply to capabilities that, if used, must conform to the standard. Spare requirements refer to capabilities that must exist in all designs, but the details on how the capability is implemented is a function of the particular design. Reserved requirements refer to capabilities that must exist in all designs but that cannot be used.

The standard calls for 16-bit fixed point, single precision; 32-bit fixed point, double precision; 32-bit floating point; and 48-bit floating point extended precision data formats. A 2's complement representation is used. In all formats the most significant bit of the first word is the sign bit.

Instructions can be either 16 or 32 bits long; 16 bits are the most common. The first 8 bits of the word contain the operation code. There are several important points to be made about 1750 instructions. Any attempt to execute an instruction for which the first 16 bits are not defined by the standard must set the invalid instruction bit in the fault register and generate a machine error interrupt. All of these undefined patterns are reserved and cannot be used.

In addition to the usual memory and register direct addressing modes, there are 12 other addressing modes. Since the ISA is based on 16-bit words, up to 65,536 words can be addressed in the direct mode.

There are 16 general registers, 5 mandatory special registers, and 3 optional special registers. Of particular importance among the mandatory registers is the fault register used to indicate machine error conditions. Twelve of the bits in this register have been assigned to indicate various types of faults (such as memory parity fail, illegal address, and illegal instruction), 3 bits are reserved (including 2 for built-in test), and there is 1 spare bit.

As stated earlier, the 16-bit words can address 65,536 words in memory. When expanded memory is required, it must be implemented through a memory paging scheme. There can be a maximum of 512 pages in the page file. These pages must be partitioned into 16 groups, and each group must contain two sets of 16-page registers each. Each page must contain 4096 words, which means that one page can address 65,536 words.

The 1750 instruction set must be capable of supporting 16 types of interrupts. Seven of the interrupts can be disabled and three cannot: power down, machine error, and executive call. The six remaining interrupts are spares. No interrupt will take effect until the currently executing instruction is completed.

Input and output commands exchange data between the central processor unit and special registers or peripherals, timers, and so on.

References

1. Frenock, Thomas J., "Advanced Integrated Flight Control System," Boeing Military Airplane Co. Contract NAS2-11866, 1985.
2. Hatley, Derek J., and Pirbhai, Imtiaz, *Strategies for Real Time System Specification*, Dorsett House, New York, 1987.
3. Gerichtin, L., et al., "Software Methodology Catalog," Technical Report C01-091JB-0001-01, U.S. Army Communications—Electronics Command, 1989.
4. Wood, David P., and Wood, William G., "Comparative Evaluations of Four Specification Methods for Real-Time Systems," Software Engineering Institute Technical Report CMU/SEI-89-TR-36, 1989.
5. DO-178A Software Considerations in Airborne Systems and Equipment Certification, RTCA, 1985.
6. DOD-STD-2167A Defense System Software Development, 29 February 1988.
7. DOD-STD-2168 Defense System Software Quality Program, 29 April 1988.
8. DOD Directive 5000.29 "Computer Programming Language Policy," 4 March 1983.
9. DOD Directive 3405.2 "Use of Ada in Weapon Systems," March 1987.
10. ARINC Report 613: Guidance for Using the Ada Programming Language in Avionics Systems, 1987.
11. Level A/B Software Management Policies for the Space Station, November 11, 1986.
12. Foreman, John, and Goodenough, John, "Ada Adoption Handbook: A Program Manager's Guide," Carnegie Mellon University Software Engineering Institute Report CMU/SEI-87-TR-9, 1987.
13. ANSI/MIL-STD-1815A Reference Manual for the Ada Programming Language (aka Language Reference Manual), 22 January 1983.
14. Westermeier, T. F., and Hansen, H. E., "The Use of Ada in a Digital Flight Control System," AIAA paper no. 85-1953-CP, 1985.
15. Bhansali, P. V., et al., "Ada Technology: Current Status and Cost Impact," *Proceedings of the IEEE*, vol. 79, no. 1, pp. 22–9, 1991.
16. Funk, David W., "Applying Ada to Beech Starship Avionics" *NASA Johnson Space Center First International Conference on Ada Programming Applications for the Space Station*, 1986.
17. Keen, W. B., "Ada in Avionics—Beyond Validation," *AIAA/IEEE Eighth Digital Avionics Systems Conference*, AIAA paper no. 88-3907-CP, 1988.
18. Humphrey, T. D., "Reducing the Risk of Using Ada Onboard the Space Station,"

IEEE Aerospace and Electronic Systems Magazine, vol. 3, no. 11; pp. 21–4, November 1988.
19. "Ada Programming Guidelines for Deterministic Storage Management" SofTech Document WO-126, prepared for Johnson Space Center, 1988.
20. MIL-STD-1750A (USAF) Sixteen-Bit Computer Instruction Set Architecture, 2 July 1980.

Bibliography

Bruyn, W. et al., "ESML: An Extended Systems Modeling Language," *ACM Software Engineering Notes,* vol. 13, no. 1; pp. 58–67, January 1988.

Harel, D., and Rolph, S., "Modeling and Analyzing Complex Reactive Systems: The Stalemate Approach," *AIAA Computers in Aerospace Conference,* 1989.

Ward, P., and Mellor, S., *Structured Development for Real-Time Systems.* Prentice-Hall, New York, Yourdon Press, 1985.

Chapter

11

Figuring the Costs of Avionics

The economic questions for avionics are a lot like those for an automobile—how much does it cost to buy, how much to operate, and what can I get for it when I sell it? This chapter explores the fundamentals of the first two of these questions as applied to digital avionics.

Aircraft operators, both civil and military, consider avionics costs to be an important element of the overall picture. The relative importance of cost, though, is different between the two. Civil operators consider cost to be as important as the performance variables such as reliability and weight. Military operators consider cost only after determining that the avionics will do their part for mission success.

As suggested in the opening paragraph of this chapter, the top-level avionics costs can be lumped into two categories: acquisition and operation. The real world, of course, is never that simple. Acquisition costs include, for example, the original equipment, initial spares and expendables, and installation costs. Operating costs include, as a minimum, maintenance personnel wages, replenishment of expendables, and maintenance support.

This chapter explores each factor in the avionics cost equation and gives enough detail to permit the designer to conduct a meaningful cost analysis. Several topics are given special attention: cash flow analysis, payback period, software costs, and determining the number of spares needed for a stated protection level.*

Some Underlying Principles

In developing the costs of digital avionics, there is generally much confusion about the precise meaning of terms. To avoid this problem,

---

*Much of the material in the following sections, Life Cycle Costs for Civil Avionics, Cash Flow Analysis for Civil Avionics, and Establishing the Spares Level, was adapted from Boeing Commercial Airplane Company Report D6-52010 Cost-Benefit Analysis Guide and Data Requirements for Suppliers (1983). Used with permission.

at least for the purposes of this book, the following terminology is established. Life cycle cost (LCC) is the total cost for a digital avionics system from conception through final disposition (see Fig. 11.1). This general definition includes all hardware and software design, development, test, production, installation, maintenance, spares, retirement costs, and tax adjustments. Civil operators divide the LCC into two parts: investment and cost of ownership. For military systems, tax adjustments and schedule interruptions shown in Fig. 11.1 are not relevant.

The designer plays a central role in determining the cost of an avionics system—a role that goes far beyond the simple impact of the cost of components on the LCC. Figure 11.2 portrays the relationship between LCC and mean time between failure (MTBF) for a typical system. Examination of this figure shows that as MTBF increases, the operating costs decrease. However, to achieve this increased MTBF requires a higher-quality design with resulting higher design, development, and testing costs. The designer must construct this pair of curves very early in the design process so that the minimum LCC design can be established. One avionics manufacturer estimates that 8000- to 10,000-MTBF leads to minimum LCC for civil transport avionics. These curves can be based on data from predecessor systems of similar design and extrapolated or modified to model the present design. The important point is to do the analysis as early as possible.

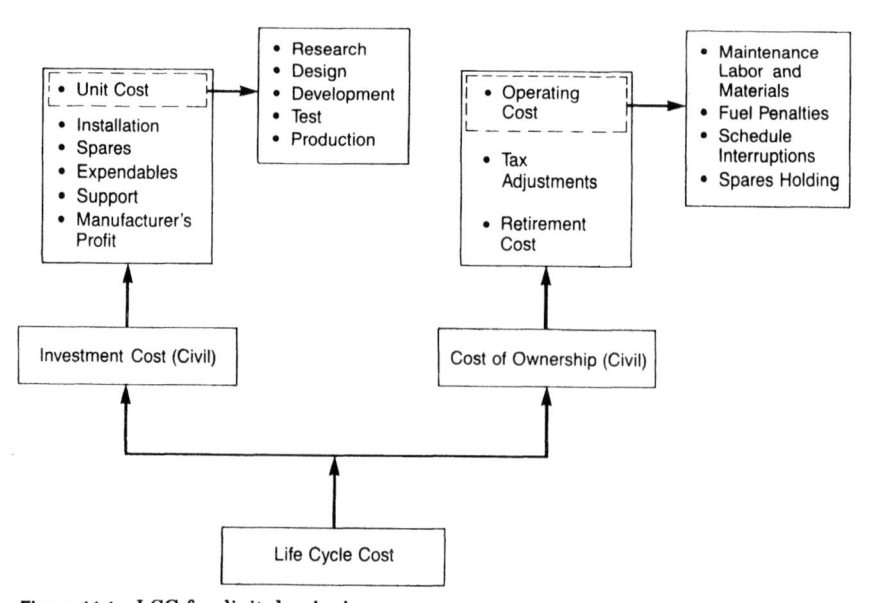

Figure 11.1 LCC for digital avionics.

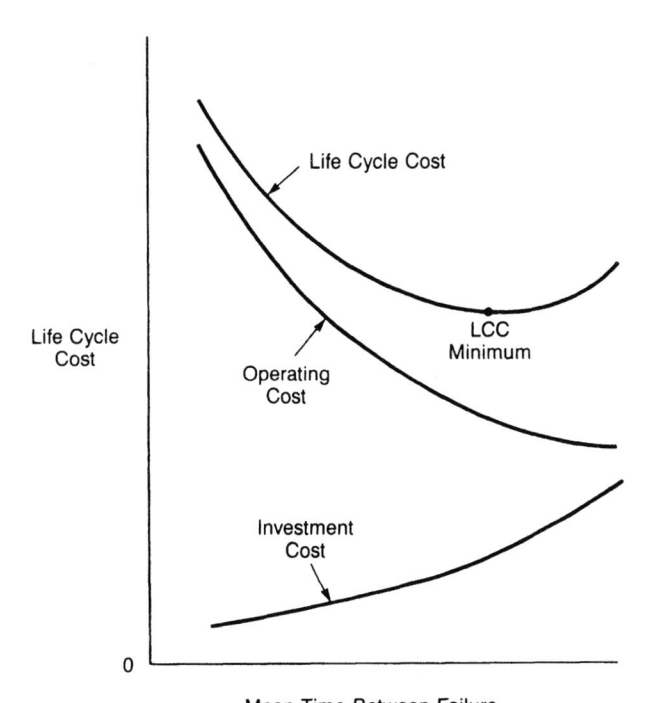

Figure 11.2 Relation of LCC to MTBF.

Life Cycle Costs for Civil Avionics

The complete LCC for civil avionics is shown in Fig. 11.1. The basic segments of LCC are investment cost, operating cost, tax adjustments, and retirement cost. The major emphasis in this section will be on the investment and operating cost segments and their elements since these are the parts of the LCC that can be significantly influenced by the designer.

The investment cost (IC) is the up-front costs for an avionics system, or what it costs to put the system into service. The elements of investment cost are fleet installed cost (FIC), fleet spares cost (FSPC), and fleet support costs (FSUC). Thus, the equation for investment cost is

$$IC = FIC + FSPC + FSUC$$
$$FIC = (UC + INS) \times QPA \times NA$$

where UC = unit cost
       INS = installation cost per unit
       QPA = quantity per aircraft
       NA = number of aircraft in fleet (for a given operator)

It is important to note that unit cost includes all of the hardware and software research, design, development, manufacturing, and testing costs pro-rated across the estimated total number of units to be produced (software costs are examined in detail in a later section of this chapter):

$$FSPC = ICRS + ICES$$

where ICRS = initial cost of rotable parts
ICES = initial cost of expendable spares

$$ICRS = UC \times NS$$

where NS = number of spares

Because of its importance and complexity, determining the NS is treated separately in the final section of this chapter.

$$ICES = MM \times NMS/12$$

where MM = maintenance material cost per year per fleet
NMS = number of months of stock in initial provisioning

Furthermore, maintenance material (MM) cost is determined by:

$$MM = (NA \times UTIL \times QPA \times FR) \times CE$$

where UTIL = flight hours per year per aircraft*
FR = failures per flight hour
CE = Cost per expendable item

As the value of MM increases, the number of months of stock (NMS) decreases to minimize the sum of reorder and inventory holding costs. Thus, the empirical relationship shown below has evolved to guide stock levels:

If	NMS is
$ 0 < MM < $ 399 (U.S. 1986 dollars)	12 months
$ 400 < MM < $ 999	6 months
$1000 < MM < $1999	4 months
$2000 < MM < $5999	2 months
$6000 < MM	1 month

*UTIL can be measured in several ways, e.g., for commercial operators it can be block time, flight time (load off land gear) or perhaps even some other time.

Thus, the equation for ICES can be rewritten as:

For $ 0 < MM < \$ 399$ ICES = MM
For $\$ 400 < MM < \$ 999$ ICES = MM × ½
For $\$1000 < MM < \$1999$ ICES = MM × ⅓
For $\$2000 < MM < \$5999$ ICES = MM × ⅙
For $\$6000 < MM$ ICES = MM × 1/12

FSUC is less rigorously defined, and the costs to be included in this element are at the discretion of the designer and the analyst. Generally, however, the costs typically included in this element are ground support and ramp equipment, test equipment, training hours and equipment, simulators, and technical documentation.

Operating cost (OC) is the largest and most important segment of LCC. The elements of OC are maintenance labor including burden (DML), direct maintenance material (DMM), fuel penalties (FP), and spares holding (SH).

$$OC = \sum_{i=0}^{N-1}\{[DML \times (1 + MBF) + DMM + FP] \times UTIL \times NA + SH\} \times (1 + INF/100)^i$$

where N = service life, years
   DML = direct maintenance labor, line and shop, dollars per flight hour
   MBF = maintenance burden factor, typically 2 < MBF < 2.4
   DMM = direct maintenance material, cost in dollars of material and expendables per flight hour
   INF = inflation rate, percent per year

$$FP + FCW + FCP$$

FCW = fuel cost in dollars per flight hour due to system weight or how much it costs to fly the weight of the system for an hour. FCW is the product of system weight, pounds of fuel per hour per pound of system weight, and the fuel price per pound.

FCP = fuel cost penalties due to energy required to operate the system, cooling air, and system induced drag penalties. FCP is usually expressed in equivalent pounds of weight and then treated the same as FCW.

Spares handling costs are typically 10 percent of the value of the rotable spares or

$$SH = 0.10 \times NS \times UC$$

Tax adjustments (TA) are the sum of investment tax credit (ITC), and income tax deductible expenses (INC):

$$TA = ITC + INC$$

The 1982 Tax Act permits the following adjustments:

ITC = IC × 0.10   (provided sufficient tax is paid to take advantage of the credit)

INC = OC × 0.48   (assuming federal and state taxes combined are approximately 48 percent.)

Retirement cost (RC) is the cost of disposition of the system and any income realized from its sale.

One very important factor not included in the equations above is operating benefits. Examples of operating benefits include increased payload, range, and ride quality and reduced weight, fuel burn, structural repair, and break even load factor. All of these benefits, with the possible exception of ride quality, can be quantitatively evaluated using equations similar to those on the previous pages. The digital avionics designer must ensure that these additional benefits are included in any economic analysis of a system design.

Cash Flow Analysis for Civil Avionics

The prior section dealt with the segments and elements of LCC for civil avionics. This section introduces additional concepts necessary for civil avionics LCC assessment and presents an illustrative LCC analysis example.

The first additional concept is inflation. The inflation rate is determined by economic conditions and, therefore, can vary widely and somewhat unpredictably during the life of a system. Thus, it is very difficult to choose a most probable future value. Including inflation rates in LCC analyses is further complicated by the fact that different elements of the LCC equation, such as labor rates and material costs, have different inflation rates. Consequently, rather than base an LCC analysis on a single inflation rate, the preferred approach is to conduct several analyses using a range of inflation rates to accommodate this uncertainty and to demonstrate LCC sensitivity to this important parameter.

The second important concept is the time value of money. The goal is to make all money, regardless of when it is expended in the life of the system, equivalent to its value at the time used in the LCC analysis (typically the time of system purchase). Thus, money spent before the time in the LCC analysis, such as progress payments, is increased to its value at the time of the analysis by the compounded minimum acceptable rate of return (MARR). (The MARR is established by the aircraft operator for all investments and is generally set several percentage points above the current or predicted corporate interest rate.) Similarly, money to be spent in the future is discounted to its value at the time in the LCC analysis by the inverse of the compounded MARR.

Armed with the equations from the prior section and the concepts described above, examples of LCC analysis can be presented.

Consider two systems: the first costs $100,000 per aircraft installed, and requires $15,000 per year in operating costs. The second system costs $90,000 installed, and requires $17,000 per year in operating costs. For both systems the following conditions apply:

- Inflation: 8 percent
- MARR: 12 percent
- System life: 15 years
- Operating benefits and retirement costs are not considered

Tables 11.1 and 11.2 show the cash flow analyses and costs of ownership for the two example systems. Table 11.3 then compares the costs of ownership. Although System 1 has a higher investment cost, it is offset by the lower operating costs, so by the fourth year of operation System 1 is the preferred system. The payback period for System 1 occurs after the fourth year when the savings in operating costs and the additional tax credits have offset the higher investment cost. At the end of 15 years, the estimated service life, System 1 will have cost $9039 less than System 2.

If operating benefits had been included in the analysis, it would have been computed on a year-by-year basis and treated the same as operating costs except it would be a positive cash flow. If retirement costs were included, they would be treated like investment cost as a one-time entry (in the fifteenth year, but discounted) and could be positive or negative depending on the cost of removing the equipment from service and its resale or salvage value.

Life Cycle Costs for Military Avionics

The concept of LCC for military avionics is very similar to that of civil avionics. The segments of LCC shown in Fig. 11.1 are essentially the same: acquisition cost, operating cost, and support cost. In military

TABLE 11.1 Simplified Cash Flow Analysis for Example System 1

Time (yr)	Tax credit	Depreciation	Operating expenses	Income tax	Total	$\frac{1}{\text{MARR}}$	Present equivalent value	Cost of ownership
0	+$10000				-$100000	1.0000	-$100000	-$100000
1		+$20000	-$15000	+$7200	+22200	0.8929	+19821	-80719
2		+32000	-16200	+7776	+23576	0.7972	+18795	-61924
3		+24000	-17496	+8398	+14902	0.7118	+10607	-51317
4		+16000	-18896	+9070	+6174	0.6355	+3924	-47393
5		+8000	-20407	+9795	-2612	0.5674	-1482	-48875
6			-22040	+10579	-11461	0.5066	-5808	-54683
7			-23803	+11425	-12378	0.4523	-5599	-60282
8			-25707	+12339	-13368	0.4039	-5399	-65681
9			-27764	+13327	-14437	0.3606	-5206	-70887
10			-29985	+14393	-15592	0.3220	-5020	-75915
11			-32384	+15544	-16840	0.2875	-4841	-80756
12			-34975	+16788	-18187	0.2567	-4668	-85424
13			-37773	+18131	-19642	0.2292	-4501	-89925
14			-40794	+19581	-21213	0.2046	-4340	-94265
15			-44058	+21148	-22910	0.1827	-4186	-98451

TABLE 11.2 Simplified Cash Flow Analysis for Example System 2

Time (yr)	Tax credit	Depreciation	Operating expenses	Income tax	Total	$\frac{1}{MARR}$	Present equivalent value	Cost of ownership
0					−$90000	1.0000	−$90000	−$90000
1	$9000	+$18,000	−$17000	+$8160	+18160	0.8929	+16215	−73785
2		+28800	−18360	+8817	+19257	0.7972	+15352	−58433
3		+21600	−19829	+9518	+11289	0.7118	+8036	−50397
4		+14400	−21415	+10279	+3274	0.6355	+2081	−48316
5		+7200	−23128	+11101	−4827	0.5674	−2739	−51055
6			−24979	+11990	−13518	0.5066	−6848	−57903
7			−26997	+12949	−14028	0.4523	−6345	−64248
8			−29135	+13985	−15150	0.4039	−6119	−70367
9			−31466	+15104	−16362	0.3606	−5900	−76267
10			−33983	+16312	−17671	0.3220	−5690	−81957
11			−36701	+17617	−19084	0.2875	−5487	−87444
12			−39638	+19026	−20612	0.2567	−5291	−92735
13			−42809	+20548	−22261	0.2292	−5092	−97827
14			−46234	+22192	−24042	0.2046	−4919	−102746
15			−49932	+23968	−25964	0.1827	−4744	−107490

TABLE 11.3 Comparison of Cost of Ownership for Example Systems 1 and 2

Time (yr)	Cost of ownership System 1	Cost of ownership System 2	Difference
0	-$100000	-$90000	+10000
1	-80719	-73785	+6934
2	-61924	-58433	+3491
3	-51317	-50397	+920
4	-47393	-48316	-923
5	-48875	-51055	-2180
6	-54683	-57903	-3220
7	-60282	-64248	-3966
8	-65681	-70367	-4686
9	-70887	-76267	-5380
10	-75915	-81957	-6042
11	-80756	-87444	-6688
12	-85424	-92735	-7311
13	-89925	-97827	-7902
14	-94265	-102746	-8481
15	-98451	-107490	-9039

avionics, operating and tax benefits and retirement costs, except equipment removal costs, are not applicable.

For military avionics the acquisition cost, similar to the investment costs for civil avionics, includes the cost of the hardware and software design, development, and test; support equipment and software; training; initial spares; installation; and checkout.

For military avionics the operation and support costs, similar to the operating costs for civil avionics, include maintenance personnel and material, spares, and spares handling. Operation costs refer primarily to energy costs and are usually considered to be negligible.

Military LCC models include separate equations for repairs performed on the aircraft, repairs performed at the base and depot levels, and provisions for line replaceable units (LRUs) that are not repairable. Additionally, the models also provide for a detailed treatment of the support equipment requirements and recurring software support.

The LCC models used by the separate military services have slight differences and are very detailed when compared to the civil LCC model discussed earlier. Thus, while there is no single "universal" military LCC model, the Standardization Evaluation Program (STEP) model, developed by a tri-service working group, is growing in capability and gaining recognition as a very powerful tool. Typical of the capabilities embedded in STEP are accounting for the learning curve in design and assembly and recognizing reliability growth.[1] STEP also

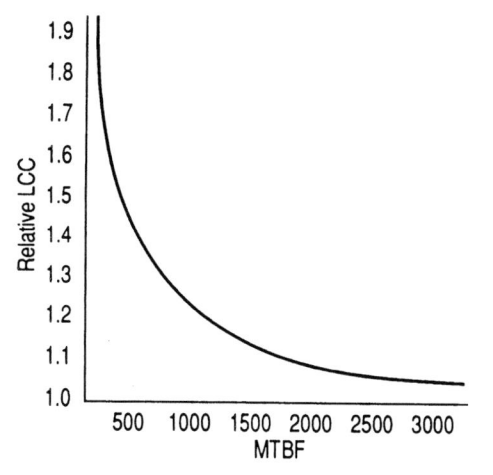

Figure 11.3 Effect of MTBF on LCC.[2]

is able to determine the relative LCC as a function of MTBF, as shown in Fig. 11.3. The curve in Fig. 11.3 duplicates and quantifies the middle curve in Fig. 11.2.[2]

Software Costs

One of the most difficult areas in avionics systems cost estimating is software. Difficulties in estimating software costs can come from imprecise specification of performance requirements, poor understanding of the requirements, and/or poor estimates of the software team capability.

As stated many times earlier in this book, it is essential that the requirements be clearly written and equally clearly understood. Given a clear understanding of the software requirements, it is possible to estimate, within reasonable limits, the number of instructions and size of the databases in the software based on prior similar situations. Where prior similar software does not exist, estimates of the program size will have to be based on expert judgment. The best possible quality of estimate of the software and database size is essential since all subsequent software costs are based on it. The more accurate the size estimate, the better the software cost estimate.

Given a good estimate of the number of lines of code and database size, the next principal uncertainty is the skill of the programmer. As will be shown later, the skill of the programmer can change the cost of a line of code by approximately a factor of 2. When developing the software cost estimate, an accurate appraisal of the skill of the team members is important.

One of the most widely used software cost estimating models is the constructive cost model (COCOMO). COCOMO has been used in a wide variety of applications that have proven its accuracy and reliability.

The COCOMO model can operate at three levels of detail from macroscopic to microscopic (that includes a complete work breakdown structure and phase sensitive multipliers). This discussion will focus on the intermediate level of COCOMO since that level is relatively easy to work with and yet yields acceptably accurate answers.

Using COCOMO is basically a two-step process. The first step is rooted in the fundamental beginning point for all software cost models: the most reliable estimate of the number of delivered source instructions and in the development mode of the project. Table 11.4 correlates eight features of the software to three possible development modes: organic, semidetached, or embedded.

With the size of the program and its development mode known, the number of programmer months and development time in months can be calculated using the equations in Table 11.5. The number of programmer months determined by the nominal effort equation in Table 11.5 is the nominal programmer months required for the complete software build process from requirement interpretation to delivery of the code.

Step 2 is to determine the effort multipliers that are applied to the nominal programmer months value to achieve the most probable number of programmer months. There are 15 cost driver attributes, shown in Table 11.6, that have varying degrees of impact on the cost of the software. Table 11.6 can be used to establish the rating for each attribute from very low to extra high, except for the complexity attribute, which is rated using Table 11.7. (Expanded rating tables for each attribute can be found in *Software Engineering Economics*.) Once the ratings are established, the effort multiplier for each attribute is extracted from Table 11.8. Assuming all 15 attributes are rated, the product of all effort multipliers can range from 0.0886 to 74.7408. Thus, it is obvious that careful attention must be given to the ratings of each of the attributes.

After the nominal programmer months value is adjusted by the effort multiplier, the schedule equation in Table 11.5 is used to estimate the time required to develop the software. As a general rule of thumb, approximately 40 percent of the time will be required for analysis and design of the software, 20 percent for coding and auditing, and 40 percent for checkout and test.

The software costs then become one element in the total avionics system cost along with design, hardware, and so on.

TABLE 11.4 COCOMO Software Development Modes*

Feature	Mode		
	Organic	Semidetached	Embedded
Organizational understanding of product objectives	Thorough	Considerable	General
Experience in working with related software systems	Extensive	Considerable	Moderate
Need for software conformance with preestablished requirements	Basic	Considerable	Full
Need for software conformance with external interface specifications	Basic	Considerable	Full
Concurrent development of associated new hardware and operational procedures	Some	Moderate	Extensive
Need for innovative data processing architectures, algorithms	Minimal	Some	Considerable
Premium on early completion	Low	Medium	High
Product size range	<50 KDSI	<300 KDSI	All sizes
Examples	Batch data reduction Scientific models Business models Familiar OS compiler Simple inventory, production control	Most transaction processing systems New OS DBMS Ambitious inventory, production control Simple command control	Large, complex transaction processing systems Ambitious, very large OS Avionics Ambitious command control

*Barry W. Boehm, *Software Engineering Economics*, © 1981. Reprinted by permission of Prentice-Hall, Inc., Englewood Cliffs, NJ.

TABLE 11.5 COCOMO Nominal Effort and Schedule Equations*

Development mode	Nominal effort	Schedule
Organic	$(MM)_{NOM} = 3.2(KDSI)^{1.05}$†	$TDEV = 2.5(MM_{DEV})^{0.38}$
Semidetached	$(MM)_{NOM} = 3.0(KDSI)^{1.12}$	$TDEV = 2.5(MM_{DEV})^{0.35}$
Embedded	$(MM)_{NOM} = 2.8(KDSI)^{1.20}$	$TDEV = 2.5(MM_{DEV})^{0.32}$

*Barry W. Boehm, *Software Engineering Economics*, © 1981. Reprinted by permission of Prentice-Hall, Inc., Englewood Cliffs, NJ.
†KDSI = thousands of delivered source instructions.

Establishing the Spares Level

One of the most important issues in introducing a digital avionics system into service is establishing the optimum number of spares. Spares have both an operational and financial impact. Operationally they are needed in the right place at the right time so that a failed unit can be replaced and the aircraft can return to flight status as quickly as possible.

As for the financial aspect, it is clear that avionics LRUs are not low-cost items, so there is pressure to keep the number, and, therefore, the cost of spares to a minimum. (In fact, when digital avionics were originally introduced, the spares level was established using estimated failure rates based on analog equivalent units. Since then, experience has shown the digital units to be much more reliable than predicted, so today the number of spare digital LRUs is being reduced.)

In addition to establishing the optimum spares level, there is the issue of how to deploy the spares. For military avionics, spares can be distributed based on the number of LRUs and LRU operating hours for each organization that has the subject LRU in its inventory. For civil aircraft, especially commercial transports, the issue is much more complex. Distribution of spares for commercial transports must consider the airline route structure (hub and spoke, point to point, spoke to spoke, and so on), the minimum equipment list (MEL), maintenance procedures, and shared or joint use of spares. For example, in a hub and spoke route structure, spares may not need to be deployed at each spoke if MEL provisions allow a flight leg to be originated to a destination (hub) that stocks the failed spare. Because deployment of spares for civil transports is a function of so many variables, this section will address only the issue of establishing the optimum number of spares from an equipment performance perspective. Guidance on the deployment of these spares comes from analysis of the individual airline operations.

TABLE 11.6 COCOMO Development Effort Ratings*

	Cost driver	Ratings					
		Very low	Low	Nominal	High	Very high	Extra high
	Product attributes						
Required software reliability	RELY	Effect: slight inconvenience	Low, easily recoverable losses	Moderate, recoverable losses	High financial loss	Risk to human life	
Database size	DATA		$\frac{DB\ bytes}{Prog.DSI} < 10$	$10 \leq \frac{D}{P} < 100$	$100 \leq \frac{D}{P} < 1000$	$\frac{D}{P} > 1000$	
Product complexity	CPLX	See Table 11.7					
	Computer attributes						
Execution time constraint	TIME			<50% use of available execution time	70%	85%	95%
Main storage constraint	STOR			<50% use of available storage	70%	85%	95%
Virtual machine volatility†	VIRT		Major change every 12 months	Major 6 months	Major 2 months	Major 2 weeks	
Computer turnaround time	TURN		Minor: 1 month Interactive	Minor 2 weeks Average turn-around <4 hours	Minor 1 week 4–12 hours	Minor 2 days >12 hours	
	Personnel attributes						
Analyst capability Applications experience	ACAP AEXP	15th percentile† <4 months experience	35th percentile 1 year	55th percentile 3 years	75th percentile 6 years	90th percentile 12 years	
Programmer capability Virtual machine experience	PCAP VEXP	15th percentile† ≤1 month experience	35th percentile 4 months	55th percentile 1 year	75th percentile 3 years	90th percentile	
Program language experience	LEXP	≤1 month experience	4 months	1 year	3 years		

TABLE 11.6 COCOMO Development Effort Ratings* (Continued)

Cost driver	Ratings					
	Very low	Low	Nominal	High	Very high	Extra high
Project attributes						
Use of modern programming practices — MODP	No use	Beginning use	Some use	General use	Routine use	
Use of software tools — TOOL	Basic microprocessor tools	Basic mini tools	Basic mini/maxi tools	Strong maxi programming, test tools	Add requirements design, management, documentation tools	
Required development schedule — SCED	75% of nominal	85%	100%	130%	160%	

*Barry W. Boehm, *Software Engineering Economics*, © 1981. Reprinted by permission of Prentice-Hall, Inc., Englewood Cliffs, NJ.
†Team rating criteria analyses (programming) ability, efficiency, ability to communicate and cooperate.

TABLE 11.7 COCOMO Module Complexity Rating versus Type of Module*

Rating	Control operations	Computational operations	Device-dependent operations	Data management operations
Very low	Straightline code with a few nonnested SP operators: DOs, CASEs, IFTHENELSEs. Simple predicates	Evaluation of simple expressions [e.g., $A = B + C (D - E)$]	Simple read, write statements with simple formats	Simple arrays in main memory
Low	Straightforward nesting of SP† operators; mostly simple predicates	Evaluation of moderate-level expressions [e.g., $D = $ SQRT $(B**2-4$ "A" $C)$]	No cognizance needed of particular processor or I/O device characteristics; I/O done at GET/PUT level; no cognizance of overlap	Single file subsetting with no data structure changes, no edits, no intermediate files
Nominal	Mostly simple nesting; some intermodule control; decision tables	Use of standard math and statistical routines; basic matrix vector operations	I/O processing includes device selection, status checking and error processing	Multifile input and single file output; simple structural changes, simple edits
High	Highly nested SP operators with many compound predicates; queue and stack control; considerable intermodule control	Basic numerical analysis (N.A.): multivariate interpolation, ordinary differential equations; basic truncation, roundoff concerns	Operations at physical I/O level (physical storage address translations, seeks, reads, etc.); optimized I/O overlap	Special-purpose subroutines activated by data stream contents; complex data rests, turning at record level
Very high	Reentrant and recursive coding; fixed-priority interrupt handling	Difficult but structured; N.A.: near-singular matrix equations, partial differential equations	Routines for interrupt diagnosis, servicing, masking; communication line handling	A generalized, parameter-driven file structuring routine; file building, command processing, search optimization
Extra high	Multiple resource scheduling with dynamically changing priorities; microcode-level control	Difficult and unstructured; N.A.: highly accurate analysis of noisy stochastic data	Device timing-dependent coding, microprogrammed operations	Highly coupled, dynamic relational structures; natural language data management

*Barry W. Boehm, *Software Engineering Economics*, © 1981. Reprinted by permission of Prentice-Hall, Inc., Englewood Cliffs, NJ.
†SP = structured programming.

TABLE 11.8 COCOMO Development Effort Multipliers*

Cost drivers	Very low	Low	Nominal	High	Very high	Extra high
Product attributes						
RELY	0.75	0.88	1.00	1.15	1.40	
DATA		0.94	1.00	1.08	1.16	
CPLX	0.70	0.85	1.00	1.15	1.30	1.65
Computer attributes						
TIME			1.00	1.11	1.30	1.66
STOR			1.00	1.06	1.21	1.56
VIRT		0.87	1.00	1.15	1.30	
TURN		0.87	1.00	1.07	1.15	
Personnel attributes						
ACAP	1.46	1.19	1.00	.86	0.71	
AEXP	1.29	1.13	1.00	.91	0.82	
PCAP	1.42	1.17	1.00	.86	0.70	
VEXP	1.21	1.10	1.00	.90		
LEXP	1.14	1.07	1.00	.95		
Project attributes						
MODP	1.24	1.10	1.00	.91	0.82	
TOOL	1.24	1.10	1.00	.91	0.83	
SCED	1.23	1.08	1.00	1.04	1.10	

*Barry W. Boehm, *Software Engineering Economics*, © 1981. Reprinted by permission of Prentice-Hall, Inc., Englewood Cliffs, NJ.
For a given software product, the underlying virtual machine is the complex of hardware and software (OS, DBMS, etc.) it calls on to accomplish its tasks.

The number of spares (NS) can be determined by the following equation if the demand for spares is assumed to match a Poisson distribution:

$$PL^{(1/K)} = \sum_{i=0}^{NS}(e^{-PI}) \times PI^i/i!$$

where PL = kit protection level
K = number of different items in the rotable spares list
PI = performance index = [(QPA × NA × UTIL)/MFHBR] × (TRD/365)
QPA = quantity per aircraft
NA = number of aircraft in fleet (for a given operator)
UTIL = flight hours per year per aircraft
MFHBR = mean flight hours between removals
TRD = turnaround time in days

Several parameters in the NS equation warrant further discussion. The PL is the probability of having *any* spare on demand from the supply of *all* rotable spares. PL is set by organization policy. However, a typical value is 0.97.

The number of different items on the rotable spares list is a function of the aircraft design. Items on the list are not limited to avionics but include *every* different type of rotable spare on the aircraft such as hydraulic pumps, actuators, compressors, and so on. A typical value for K is 600.

The quantity, (QPA × NA × UTIL)/MFHBR, is the total number of removals of the unit under analysis from all aircraft in a given operator's fleet for 1 year.

The MFHBR is not the same as the mean flight hours between failures. Removals are more frequent than failures since some removed equipment will retest as satisfactory during troubleshooting and no failure can be found. The degree to which these two quantities match is a measure of the quality of the avionics built-in test equipment. Minimize unnecessary removals.

The number of aircraft in the fleet (NA) is for a given operator, not the total number of aircraft of that type.

TRD is the number of days from when a unit is removed from ser-

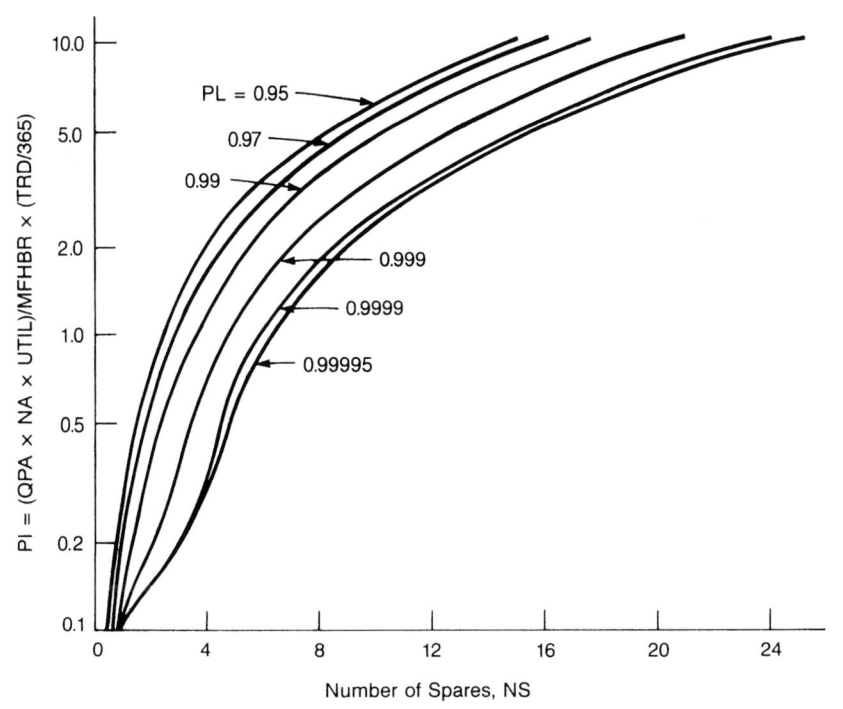

Figure 11.4 Number of spares (NS) as a function of the performance index (PI) for selected kit protection levels (PL).

vice until it is in the spare queue ready for use again. TRD is typically 12 days: 5 days in transit and 7 days for repair.

Figure 11.4 is a plot of the spares equation as a function of the PI. The figure shows the intuitive results: As operating hours and turnaround time increase, the number of spares for a given protection level increases, and as the removal rate decreases, the number of spares also decreases.

Particular attention should be given to the repair turnaround time since a reduction there can yield a substantial reduction in the number of spares required. The reduction in spares investment cost will very likely more than offset the additional cost of reducing repair times.

References

1. Dickman, Thomas J., and Roberts, Thomas M., "Modular Avionics System Architecture Decision Support System." *Proceedings of the 1989 National Aerospace Electronics Conference,* IEEE 89CH2759-9. pp. 1549–52, 1989.
2. Curry, Ernest E., "STEP: A Tool for Estimating Avionics Life Cycle Costs," *IEEE Aerospace and Electronic Systems Magazine,* January 1989, pp. 30–2, 1989.

Glossary

ACARS Aircraft Communications Addressing and Reporting System.

Accident (aircraft): An occurrence associated with the operation of an aircraft in which the aircraft receives substantial damage (FAR 830.2). (Substantial damage means damage or structural failure which adversely affects the structural strength, performance, or flight characteristics of the aircraft and which would normally require major repair or replacement of the affected component.)

Ada A higher-order programming language that must be used in embedded military avionics computers (ANSI/MIL-STD-1815).

AFTI Advanced Fighter Technology Integration, a U.S. Air Force high-performance aircraft research program.

AIAA American Institute of Astronautics and Aeronautics.

Air transport racking see Austin Trumbull Radio.

Allocated baseline The initially approved documentation describing an item's functional and interface characteristics that are allocated from those of a higher-level configuration item, interface requirements with interfacing configuration items, additional design constraints, and the verification required to demonstrate the achievement of those specified functional and interface characteristics (MIL-STD-480B).

Analytical redundancy Software to compute the most reasonable value expected from a failed sensor by algorithmically combining values from remaining functioning sensors. The computed value is treated as valid data by all users.

ANSI American National Standards Institute.

ARIES Automated Reliability Interactive Estimation System, a computer-based system reliability prediction program.

ARINC Aeronautical Radio, Inc.

ASSIST Abstract Semi-Markov Specification Interface to the SURE Tool.

ATE Automatic test equipment.

ATLAS Abbreviated Test Language for All Systems, a standard, abbreviated English language used in the preparation and documentation of test specifications, which can be implemented either manually or with automatic or semiautomatic test equipment (ANSI/IEEE Std 416).

ATR (Austin Trumbull Radio) A standard design of enclosures, shelves, cooling air, and connectors for line replaceable units of avionics. The width of a 1-ATR-wide line replaceable unit (LRU) is 10.09 in. The standard ATR height is 7.64 in and the standard depth is 12.76 in. 1 ATR = 8 modular concept units (MCUs) (ARINC Specification 600).

Avionics All the electronic and electromechanical systems and subsystems installed in an aircraft or attached to it. Does not include power generation systems.

Baseline The documented, approved description of the system at a specific time (RTCA DO-178).

BCD Binary coded decimal.

BIT Built-in test.

BITE Built-in test equipment.

BNR 2's complement fractional binary notation.

Body Implementation details of a program unit in an Ada program (ANSI/MIL-STD-1815).

Brickwalled See partitioning.

C/ATLAS Combined Abbreviated Test Language for All Systems (IEEE Std 716).

CA *See* criticality analysis.

Candela Same as one candle. *See* candle.

Candle One lumen per steradian.

CARE III Computer Aided Reliability Estimation, Version III, a computer-based system reliability prediction program.

Category I An instrument approach procedure which provides for approaches to a decision height (DH) of not less than 200 ft with a 200-ft or greater ceiling and visibility of not less than ½ mile or runway visual range (RVR) 2500 ft (RVR 1800 ft with operative touchdown zone and runway centerline lights) (FAA Advisory Circular 120-29).

Category II An instrument approach procedure which provides approaches to minima of less than decision height (DH) 200 ft/RVR 2400 ft to as low as DH 100 ft/RVR 1200 ft with a ceiling of 100 ft or greater (FAA Advisory Circular 120-29).

Category IIIa Operations with no decision height limitation to and along the surface of the runway with external visual reference during the final phase of the landing and with RVR not less than 700 ft and a ceiling of 50 ft or greater.

Category IIIb Operations with no decision height limitation to and along the surface of the runway with runway visual range not less than 150 ft, a ceiling of 35 ft or greater, and with reliance on the system for part or all of the rollout along the runway and with external visual reference for guidance along the taxiway.

Category IIIc Operations with no ceiling or decision height limitation to and along the surface of the runway and taxiways without reliance on external visual reference.

Centralized architecture A system design characterized by signal conditioning and computations taking place in one computer or several very tightly coupled computers in (usually) one line replaceable unit located in the avionics bay, with signals transmitted over one-way data buses.

Critical Systems or functions with a probability of failure of less than 10^{-9} per flight hour. This probability of failure is designated "extremely improbable" (FAA Advisory Circular 25.1309-1A).

Criticality analysis (1) An analysis to establish the criticality of a failure mode by combining its criticality classification as determined by a failure mode and effects analysis with its probability of occurrence (MIL-STD-1629); (2) an analysis conducted using service experience, engineering, operational judgment, or by using a top-down deductive qualitative analysis which examines each function performed by the system (FAA Advisory Circular 25.1309-1).

CRT Cathode ray tube.

DATAC *See* Digital autonomous terminal access communication.

Digital autonomous terminal access communication A carrier sense, multiple access, collision avoidance two-way data bus concept with an operating frequency and word structure identical to MIL-STD-1553 data buses but without a bus controller. Now called ARINC 629.

Direct operating cost (DOC) The cost of depreciation, interest, insurance, flight crew, maintenance, fuel, and oil for civil transport aircraft. Usually expressed as a function of time or distance or per flight.

Distributed architecture A system design characterized by multiple processors throughout the aircraft assigned computing tasks on a real-time basis as a function of mission phase and/or system status, and processing also is performed at the sensors and actuators.

DOC *See* direct operating cost.

DOD Department of Defense.

EL Electroluminescence.

Electromagnetic interference Any electromagnetic energy which interrupts, obstructs, or otherwise degrades or limits the effective performance of telecommunications (communication-electronic) equipment (MIL-STD-463A).

EMI *See* Electromagnetic interference.

Error A mistake in specification, design, or implementation of software (RTCA DO-178).

Essential (1) A function which the loss of degrades the flight control system performance beyond Operational State III (MIL-F-9490); (2) a system or function with a probability of failure in the range from 10^{-9} to 10^{-5} per flight hour. This probability of failure is designated "improbable" (FAA Advisory Circular 25.1309-1A).

Extremely improbable Probability of failure of a system or function of less than 10^{-9} per flight hour. Systems or functions that require this level of reliability are designated "critical" (FAA Advisory Circular 25.1309-1A).

Extremely remote The probability of failure in a flight control system (MIL-F-9490), specifically:

Heavy bombers, transports, cargo, and tankers:	$<= 5 \times 10^{-7}$/h
Rotorcraft:	$<= 25 \times 10^{-7}$/h
All other aircraft:	$<= 100 \times 10^{-7}$/h

FAA Federal Aviation Administration.

FAR Federal Aviation Regulation.

Failure (1) the lowest identifiable level of abnormal occurrence (NUREG 0492); (2) the effect of a fault (RTCA DO-178); (3) a physical event, causes a fault (Avizienis, 1975).

Failure modes and effects analysis A functionally oriented analysis of a system leading from the specific to the general case (e.g., given a malfunction of a specific device, what are the potential effects on the total system?).

Fault (1) A hardware or software error or defect in specification, design, implementation, operation, or maintenance that may lead to a failure; (2) a higher order event that can be broken down into basic events known as failures (NUREG 0492); (3) the activation of an error (RTCA DO-178); (4) incorrect input(s) to the computing process (and may cause an error) (*see* Failure 3) (Avizienis, 1975).

Fault tolerance (1) The ability to continue satisfactory operation in the presence of one or more nonsimultaneously occurring faults; (2) the built-in capability of a system to provide continued correct execution in the presence of a limited number of hardware or software faults (IEEE in ARINC 651).

Fault tree analysis (1) A functionally oriented, deductive analysis of a system leading from the general to the specific case (e.g., given an event such as a system malfunction, what are the possible causes?); (2) an analytical technique whereby an undesired state of the system is specified (usually a state that is critical from a safety standpoint) and the system is then analyzed in the context of its environment and operation to find all credible ways in which the undesired event can occur (NUREG 0492).

Federated architecture A system design characterized by each major system, such as thrust management or flight management, sharing input and sensor data from a common set of hardware and subsequently sharing their computed results over data buses.

Flight phase essential Same as essential (definition 1), except it applies only during specific flight phases (MIL-F-9490).

FMEA *See* failure modes and effects analysis.

FTA *See* fault tree analysis.

Functional baseline The initially approved documentation describing a system's or item's functional characteristics and the verification required to demonstrate the achievement of those specified functional characteristics (MIL-STD-480B).

Hardened kernel A small, simple, completely validated, minimum-capability program that is invoked in case of failure of a more complex, versatile primary program.

Host computer Any computer used to develop software for another (target) computer (RTCA DO-178).

HUD Head-up display.

ICAO International Civil Aviation Organization.

IEEE Institute of Electrical and Electronics Engineers.

Improbable Probability of failure of a system or function in the range from 10^{-9} to 10^{-5} per flight hour. Systems or functions that require this level of reliability are designated "essential" (FAA Advisory Circular 25.1309-1A).

Incident (aircraft) Flight control system malfunction or failure (FAR 830.2).

Instruction set architecture The specification of the interface between the software and the hardware (MIL-STD-1750).

ISA *See* instruction set architecture.

ISO International Standards Organization.

LCC *See* life cycle cost.

LCD Liquid crystal display.

LED Light-emitting diode.

Life cycle cost The total cost of a system throughout its lifetime including design, development, production, operation, maintenance, logistics, and disposition costs.

LRU Line replaceable unit, a single stand-alone unit with a unique part number that can be installed or removed from an aircraft by line maintenance personnel while the aircraft is on the flight line.

Lumen The amount of luminous flux in a solid angle of 1 steradian radiating from $\frac{1}{60}$ cm^2 of platinum at a temperature of 2046.65°C.

Luminous efficiency Luminous energy per unit electrical energy.

Luminous energy Radiated energy capable of evoking response in the human eye.

Maintainability The measure of the ability of an item to be retained in or restored to specified condition when maintenance is performed by personnel having specified skill levels, using prescribed procedures and resources, at each prescribed level of maintenance and repair (MIL-STD-721C).

Manchester A bi-phase method of encoding binary digits. A binary 1 begins with a positive voltage and swings negative at mid-bit. A binary 0 begins with a negative voltage and swings positive at mid-bit.

Mesh A bus configuration with multiple transmitters and receivers connected through nodes to wire or optical communication networks. Different paths can be selected as dictated by the current system configuration or in case of failure.

MCU *See* modular concept unit.

MFK Multifunction keyboard.

Minimum cut set A smallest combination of events which, if they all occur (under prescribed conditions or sequences, if any), will cause a top-level event to occur (NUREG 0492).

Modular concept unit (MCU) A designation of the size of a line replaceable unit (LRU). A 1-MCU-wide LRU is 1.00 in wide, 7.64 in tall, and 12.76 in deep. An 8-MCU LRU is 10.09 in wide. 8 MCU = 1 ATR.

Module (1) A uniquely identified element of a computer program which performs a specific function or set of related functions (RTCA DO-178); (2) limited aggregates of data and contiguous code that perform identifiable functions (DOD-STD-2167).

NASA National Aeronautics and Space Administration.

N-version programming Two or more versions of a program designed to perform the same function but prepared by independent, separate programming teams. The different versions may operate in different types of processors.

Noncritical Loss of function does not affect flight safety or reduce control capability beyond that required for Operational State III (MIL-F-9490).

Nonessential Systems or functions with a probability of failure greater than 10^{-5} per flight hour. This probability of failure is designated "probable" (FAA Advisory Circular 25.1309-1A).

NUREG Nuclear Regulation (published by the Nuclear Regulatory Commission).

One-way bus A data bus with one transmitter and one or more receivers on a single pair of wires or optical fiber.

Package The basic unit which describes logically related entities in an Ada program (ANSI/MIL-STD-1815).

Partitioning Limiting a failure to the subsystem in which it occurred, and the effects of the failure are not allowed to cascade to the rest of the system.

Probable Probability of failure greater than 10^{-5} per flight hour. Systems or functions with this probability of failure are designated "nonessential" (FAA Advisory Circular 25.1309-1A).

Product baseline The initially approved documentation describing all of the necessary functional and physical characteristics of the configuration item, any required joint and combined operation interoperability characteristics of a configuration item (including a comprehensive summary of the other service(s) and allied interfacing configuration items or systems and equipments), and the selected functional and physical characteristics designated for production acceptance testing and tests necessary for support of the configuration item (MIL-STD-480B).

Program unit The basic element of an Ada program. Program units may be subprograms, packages, or tasks. (ANSI/MIL-STD-1815).

Recovery block An alternate version of a program that is invoked if the results achieved by the primary program do not meet acceptability criteria.

Reliability (1) The duration or probability of failure-free performance under stated conditions (MIL-STD-721C); (2) the probability that an item can perform its intended function for a specified interval under stated conditions. (For nonredundant items this is equivalent to definition 1. For redundant items this is equivalent to the definition of mission reliability.) (MIL-STD-721C); (3) the property of an avionics system or component to perform over a predictable time period; usually expressed as mean time between failure (MTBF) (ARINC 651).

RTCA Formerly Radio Technical Commission for Aeronautics. Now has no specific meaning.

Specification (1) Information required by other units in an Ada program (ANSI/MIL-STD-1815); (2) engineering definition for materiel and practices resulting from new engineering and operational systems development and for items authorized for production or support (DOD Guide 4120.3).

Standard Basic design criteria, engineering nomenclature, and uniform data management practices—to assure the required reliability, maintainability, interchangeability, and compatibility (DOD Guide 4120.3).

Star A bus with a central node through which all signals pass. Used with optical media.

Subprogram Algorithms that describe either procedures or functions in an Ada program (ANSI/MIL-STD-1815).

SURE Semi-Markov Unreliability Range Estimator.

Survivability Capability of a system to continue to function in the presence of a nonnuclear threat (MIL-STD-2069).

Susceptibility A measure of the probability that an object will be hit by a given threat (MIL-STD-2069).

Tailoring The process of choosing or altering test procedures, conditions, values, tolerances, measures of failure, and so on, to simulate or exaggerate the effects of one or more forcing functions (MIL-STD-810).

Target computer The computer on which the software under development is intended to operate (RTCA DO-178).

Task unit A sequence of actions which may occur in parallel with other tasks in an Ada program (ANSI/MIL-STD-1815).

Two-way bus A data bus with multiple transmitters and receivers that can send and receive data over a single pair of wires or an optical fiber.

Unit The smallest logical entity specified in the detailed software design which completely describes a single function in sufficient detail to allow implementing code to be produced and tested independently of other units. Units are the actual entities implemented in code (DOD-STD-2167).

USAF United States Air Force.

UUT Unit under test.

Validation (1) The process of establishing that the product, of which the software is a part, complies with equipment, system, or aircraft level requirements (RTCA DO-178); (2) the process of evaluating software to determine compliance with specified requirements (DOD-STD-2167A); (3) the process of evaluating software at the end of its development cycle to ensure that it complies with its requirements, that is, evaluating whether the right software product was built (IEEE Std 727-1983).

Verification (1) The process of establishing that the software satisfies its requirements (RTCA DO-178); (2) the process of evaluating the products of a given software development activity to determine correctness and consistency with respect to the products and standards provided as inputs to that activity (DOD-STD-2167A); (3a) the process of determining whether the products of a given phase of the software development cycle fulfill the requirements established during the previous phase, (b) formal proof of program correctness, (c) the act of reviewing, inspecting, testing, checking, auditing, or otherwise establishing and documenting whether items, processes, services, or documents conform to specified requirements (IEEE in ARINC 651); (4) the process of determining whether or not the products of a given development phase fulfill the requirements established during the preceding development phase, that is, determining if the products are being built correctly (IEEE Std 727-1983).

Vulnerability A measure of the characteristics that contribute to the degradation or loss of a function (MIL-STD-2069).

Weapon replaceable assembly Alternate designation for line replaceable unit (LRU) used by the U.S. Navy. *See* LRU.

Appendix A

Environmental Testing

Part A: DO-160 Tests*

Temperature and altitude

These tests determine the performance characteristics of the equipment at the applicable categories for the temperatures and altitudes specified in Table A-1 and at the pressures defined in Table A-2.

Temperature variation

This test determines performance characteristics of the equipment during normal temperature variations between high and low operating temperature extremes specified for the applicable categories of Table A-1 during flight operations. This is a dynamic test, and it is required that the equipment be subjected to this temperature variation test when such equipment is tested according to the procedures contained in the temperature and altitude test above.

Humidity

This test determines the ability of the equipment to withstand either natural or induced humid atmospheres. The main adverse effects to be anticipated are:

1. Corrosion
2. Change of equipment characteristics resulting from the absorption of humidity, for example:

---

*DO-160C Environmental Conditions and Test Procedures for Airborne Equipment. Radio Technical Commission for Aeronautics, 1989.

TABLE A.1 Temperature and Altitude Criteria

Environmental Tests	A 1	A 2	A 3	B 1	B 2	B 3	B 4	C 1	C 2	C 3	C 4	D 1	D 2	D 3	E 1	E 2	F 1	F 2	F 3
Low operating temp., °C	−15	−15	−15	−20	−45	−45	Note (4)	−20	−55	−55	Note (4)	−20	−55	−55	−55	−55	−20	−55	−55
High operating temp., °C	+55	+70	+70	+55	+70	Note (3)	Note (4)	+55	+70	Note (3)	Note (4)	+55	+70	Note (3)	Note (3)	Note (3)	+55	+70	Note (3)
High short-time operating temp., °C	+70	+70	+85	+70	+70	Note (3)	Note (4)	+70	+70	Note (3)	Note (4)	+70	+70	Note (3)	Note (3)	Note (3)	+70	+70	Note (3)
Loss of cooling test, °C	+30	+40	+45	+30	+40	Note (3)	Note (3)	+30	+40	Note (3)	Note (3)	+30	+40	Note (3)	Note (3)	Note (3)	+30	+40	Note (3)
Low temperature ground survival, °C	−55	−55	−55	−55	−55	Note (3)	−55	−55	−55	Note (3)	−55	−55	−55	−55	−55	−55	−55	−55	−55
High temperature ground survival, °C	+85	+85	+85	+85	+85	Note (3)	+85	+85	+85	Note (3)	+85	+85	+85	Note (3)	+85	Note (3)	+85	+85	Note (3)
Altitude Thousands of feet / Thousands of meters	15 / 4.6	15 / 4.6	15 / 4.6	25 / 7.6	25 / 7.6	25 / 7.6	25 / 7.6	35 / 10.7	35 / 10.7	35 / 10.7	35 / 10.7	50 / 15.2	50 / 15.2	50 / 15.2	70 / 21.3	70 / 21.3	55 / 16.8	55 / 16.8	55 / 16.8
Decompression test	Note (1) (4)	Note (1) (4)	Note (1) (4)	—	—	—	—	—	—	—	—	—	—	—	—	—	—	—	—
Overpressure test	Note (2)	Note (2)	Note (2)	—	—	—	—	—	—	—	—	—	—	—	—	—	—	—	—

Notes: (1) The lowest pressure applicable for the decompression test is the maximum operating altitude for the aircraft in which the equipment will be installed. (2) The absolute pressure is 170 kPa (−15,000 ft, or −4600 m). (3) To be declared by the equipment manufacturer relative to temperature extremes. (4) To be declared by the equipment manufacturer and defined in the manufacturer's installation instruction when specific critical criteria exist.

TABLE A.2 Pressure Values for Various Pressure Altitude Levels

		Absolute pressure		
Pressure altitude	kPa	mbars	in Hg	mm Hg
−15,000 ft (−4,572 m)	169.73	1697.3	50.12	1273.0
−1,500 ft (−457 m)	106.94	1069.4	31.58	802.1
0 ft (0 m)	101.32	1013.2	29.92	760.0
+8,000 ft (+2,438 m)	75.26	752.6	22.22	564.4
+15,000 ft (+4,572 m)	57.18	571.8	16.89	429.0
+25,000 ft (+7,620 m)	37.60	376.0	11.10	282.0
+35,000 ft (+10,668 m)	23.84	238.4	7.04	178.8
+50,000 ft (+15,240 m)	11.60	116.0	3.42	87.0
+55,000 ft (+16,764 m)	9.12	91.2	2.69	68.3
+70,000 ft (+21,336 m)	4.44	44.4	1.31	33.3

Mechanical (metals)
Electrical (conductors and insulators)
Chemical (hygroscopic elements)
Thermal (insulators)

Operational shocks and crash safety

The operational shock test verifies that the equipment will continue to function within performance standards after exposure to shocks experienced during normal aircraft operations. These shocks may occur during taxiing or landing or when the aircraft encounters sudden gusts in flight.

The crash safety test verifies that certain equipment will not detach from its mountings or separate in a manner that presents a hazard during an emergency landing. It applies to equipment installed in compartments and other areas of the aircraft where equipment detached during emergency landing could present a hazard to occupants, fuel systems, or emergency evacuation equipment.

Vibration

These tests demonstrate that the equipment complies with the applicable equipment standards when subjected to vibration in the normal operational environment.

Explosion proofness

This test specifies requirements and procedures for aircraft equipment which may come into contact with flammable fluids and vapors. It also refers to normal and fault conditions that could occur in areas that are or may be subjected to flammable fluids and vapors during flight operations.

The flammable test fluids, vapors, or gases referred to in this section simulate those normally used in conventional aircraft and which require oxygen for combustion (for example, monofuels are not included).

These standards do not relate to potentially dangerous environments occurring as a result of leakage from goods carried on the aircraft as baggage or cargo.

In the order of testing, it is assumed that the article being tested has been subjected to the other environments of this document, as appropriate, prior to this test (i.e., altitude tests, vibration tests, etc.).

Waterproofness

These tests determine whether the equipment can withstand the effects of liquid water being sprayed or falling on the equipment.

Fluids susceptibility

These tests demonstrate whether the materials used in the construction of the equipment can withstand the deleterious effects of fluid contaminants.

Fluids susceptibility tests should only be performed when the equipment will be installed in areas where fluid contamination could be commonly encountered. The fluids are representative of those commonly used fluids encountered in airborne and ground operations. Fluids not listed herein and for which susceptibility tests are indicated shall be included in the relevant equipment specification.

Sand and dust

This test determines the resistance of the equipment to the effects of blowing sand and dust where carried by air movements at moderate speeds. The main adverse effects to be anticipated are:

1. Penetration into cracks, crevices, bearings, and joints, causing fouling and/or clogging of moving parts, relays, filters, etc.
2. Formation of electrically conductive bridges
3. Action as nucleus for the collection of water vapor, including secondary effects of possible corrosion
4. Pollution of fluids

Fungus resistance

These tests determine whether equipment material is adversely affected by fungi under conditions favorable for their development, namely, high humidity, warm atmosphere, and presence of inorganic salts.

Salt spray

This test either determines the effects on the equipment of prolonged exposure to a salt atmosphere or to salt spray experienced in normal operations. The main adverse effects to be anticipated are:

1. Corrosion of metals
2. Clogging or binding of moving parts as a result of salt deposits
3. Insulation fault
4. Damage to contactors and uncoated wiring

Magnetic effect

This test determines the magnetic effect of the equipment to assist the installer in choosing the proper location of the equipment in the aircraft.

Power input

This test defines conditions of electrical power applied to the equipment under test and enumerates the related equipment test procedures where applicable.

The equipment categories* are as follows:

Category A: Equipment intended for use on aircraft electrical systems where the primary power is from constant frequency ac generators and where the dc system is supplied from transformer-rectifier units is identified as Category A. A battery may be floating on the dc bus.

Category B: Equipment intended for use on aircraft electrical systems supplied by engine-driven alternator/rectifiers or dc generators where a battery of significant capacity is floating on the dc bus at all times is identified as Category B.

---

*When equipment requires only dc input power and is tested to the dc input parameters of Category A, B, or Z, the appropriate category shall be designated on the environmental qualification form.

Category E: When equipment requires only ac input power and is tested to the ac input parameters, the equipment is identified as Category E.

Category Z: Equipment which may be used on all other types of aircraft electrical systems applicable to these standards is identified as Category Z. Category Z shall be acceptable for use in lieu of Category A. Examples of this category are dc systems supplied from variable-speed range generators where:

> The dc power supply does not have a battery floating on the dc bus or the
>
> Control or protective equipment may disconnect the battery from the dc bus, or the
>
> Battery capacity is small compared with the capacity of the dc generators

Voltage spike

This test determines whether the equipment can withstand the effects of voltage spikes arriving at the equipment on its power leads, either ac or dc. The main adverse effects to be anticipated are:

1. Permanent damage, component failure, insulation breakdown
2. Susceptibility degradation or changes in equipment performance

The equipment categories are:

Category A: Equipment intended primarily for installation where a high degree of protection against damage by voltage spikes is required is identified as Category A.

Category B: Equipment intended primarily for installations where a lower standard of protection against voltage spikes is acceptable is identified as Category B. *Note:* For this test, equipment which derives ac power from an inverter provided exclusively for the equipment shall be considered as dc operated.

Audio frequency conducted susceptibility—power inputs (closed circuit test)

This test determines whether the equipment will accept frequency components of a magnitude normally expected when the equipment is installed in the aircraft. These frequency components are normally harmonically related to the power source fundamental frequency.

The equipment categories are:

Category A: Equipment intended for use on aircraft electrical systems where the primary power is from constant frequency ac generators and where the dc system is supplied from transformer-rectifier units is identified as Category A. A battery may be floating on the dc bus.

Category B: Equipment intended for use on aircraft electrical systems supplied by engine-driven alternator/rectifiers or dc generators where a battery of significant capacity is floating on the dc bus at all times is identified as Category B.

Category E: When equipment requires only ac input power and is tested to the ac input parameters, the equipment is identified as Category E.

Category Z: Equipment which may be used on all other types of aircraft electrical systems applicable to these standards is identified as Category Z. Category Z shall be acceptable for use in place of Category A. Examples of this category are dc systems supplied from variable-speed range generators where:

The dc power supply does not have a battery floating on the dc bus, or

Control or protective equipment may disconnect the battery from the dc bus, or

The battery capacity is small compared with the capacity of the dc generators

Induced signal susceptibility

This test determines whether the equipment interconnect circuit configuration will accept a level of induced voltages caused by the installation environment. This section relates specifically to audio frequency signals and transients that are generated by other on-board equipment.

The equipment categories are:

Category Z: Equipment intended primarily for operation in systems where interference-free operation is required is identified as Category Z.

Category A: Equipment intended primarily for operation where interference-free operation is desirable is identified as Category A.

Category B: Equipment intended primarily for operation in systems where interference should be controlled to a tolerable level is identified as Category B.

Radio frequency susceptibility (radiated and conducted)*

These tests determine whether equipment will operate within performance specifications when the equipment and its interconnecting wiring are exposed to levels of (radio frequency) RF modulated power, either by a radiated RF field or by injection probe induction onto the power lines and interface circuit wiring.

The equipment categories are:

Categories W and Y: Equipment and interconnecting wiring installed in severe electromagnetic environments. Such environments might be found in nonmetallic aircraft or exposed areas in metallic aircraft.

Category V: Equipment and interconnecting wiring installed in a moderate environment such as more electromagnetically open areas of an aircraft composed principally of metal.

Category U: Equipment and interconnecting wiring installed in a partially protected environment such as an avionics bay in an all-metallic aircraft.

Category T: Equipment and interconnecting wiring installed in a well-protected environment such as an enclosed avionics bay in an all-metallic aircraft.

Category X: Equipment and interconnecting wiring for which electromagnetic effects are insignificant or not applicable.

Emission of radio frequency energy

These tests determine that the equipment does not emit undesired RF noise in excess of the levels specified. *Note:* The problem of describing the gross RF interference environment (RF conducted and radiated susceptibility tests) in an aircraft is inseparably related to the delineation of the maximum level of spurious RF energy that any one electrical or electronic equipment in that aircraft will emit. Therefore, if

---

*See the emission of radio frequency energy test for information on the relationship between the emission of spurious radio frequency energy from electrical and electronic equipment installed in an aircraft and the levels of radio frequency susceptibility signals used in this test procedure.

the RF conducted and radiated susceptibility tests are to achieve their intended purposes, a compatible standard on the maximum permissible level of spurious emission of radio frequency energy from any one electrical/electronic equipment or instrument in an aircraft must be applied to that equipment.

The equipment categories are:

Category Z: Equipment intended primarily for operation in systems where interference-free operation is required is identified as Category Z.

Category A: Equipment intended primarily for operation where interference-free operation is desirable is identified as Category A.

Category B: Equipment intended primarily for operation in systems where interference should be controlled to a tolerable level is identified as Category B.

Lightning induced transient susceptibility

These tests will determine the ability of equipment to withstand the induced effects of lightning as defined in the individual equipment specification.

The equipment categories are:

Category J: Equipment and interconnecting wiring that will be installed in a partially protected environment such as an enclosed avionics bay in an all-metallic aircraft.

Category K: Equipment and interconnecting wiring that will be installed in a moderate environment such as the more electromagnetically open areas (e.g., cockpit) of an aircraft composed principally of metal.

Category L: Equipment and interconnecting wiring that will be installed in severe electromagnetic environments. Such levels might be found in all-composite aircraft or other exposed areas in metallic aircraft.

Category X: Equipment for which lightning effects are insignificant or not applicable.

Icing

These tests determine performance characteristics for equipment that must operate when exposed to icing conditions that would be encountered under conditions of rapid changes in temperature, altitude, and humidity.

Part B: MIL-STD-810E Tests*

Low pressure

This test determines if materiel can withstand and operate in a low-pressure environment.

High temperature

This test determines if materiel can be stored and operated under hot climatic conditions without experiencing physical damage or deterioration in performance.

Temperature shock

This test determines if materiel can withstand sudden changes in the temperature of the surrounding atmosphere without experiencing physical damage or deterioration in performance.

Low temperature

This test determines if materiel can be stored, manipulated, and operated under pertinent low-temperature conditions without experiencing physical damage or deterioration in performance.

Solar radiation (sunshine)

This test determines the effects of solar radiation on equipment that may be exposed to sunshine during operation or unsheltered storage on the earth's surface or in the lower atmosphere.

Rain

This test determines the following:

1. The effectiveness of protective covers or cases in preventing the penetration of rain
2. The capability of the test item to satisfy its performance requirements during and after exposure to rain
3. The physical deterioration of the test item caused by the rain

---

*MIL-STD-810E Environmental Test Methods and Engineering Guidelines, 9 February 1990.

Humidity

This test determines the resistance of materiel to the effects of a warm, humid atmosphere.

Fungus

This test assesses the extent to which the test item will support fungal growth or how the fungal growth may affect performance or use of the test item.

Salt fog

Salt fog climatic chamber tests determine the resistance of equipment to the effects of an aqueous salt atmosphere.

Sand and dust

This test is divided into two procedures. The small-particle procedure (dust, fine sand) is performed to ascertain the ability of equipment to resist the effects of dust particles which may penetrate into cracks, crevices, bearings, and joints. The blowing sand test is performed to determine whether materiel can be stored and operated under blowing sand (149 to 850 μm particle size) conditions without experiencing degradation of its performance, effectiveness, reliability, and maintainability due to the abrasion (erosion) or clogging effect of large, sharp-edged particles.

Explosive atmosphere

This test demonstrates the ability of equipment to operate in flammable atmospheres without causing an explosion or to prove that a flame reaction occurring within an encased equipment will be contained and will not propagate outside the test item.

Leakage (immersion)

This test determines whether materiel is constructed so that it can be immersed in water without leakage of the water into the enclosure.

Acceleration

This test assures that equipment can structurally withstand the g forces that are expected to be induced by acceleration in the service environment and can function without degradation during and following exposure to these forces.

Vibration

This test determines the resistance of equipment to vibrational stresses expected in its shipment and application environments.

Acoustic noise

This test measures how well a piece of equipment will withstand or operate in intense acoustic noise fields. The acoustic noise test complements tests for structure-borne vibrations.

Shock

This test assures that materiel can withstand the relatively infrequent nonrepetitive shocks or transient vibrations encountered in handling, transportation, and service environments. Shock tests are also used to measure an item's fragility so that packaging may be designed to protect it, if necessary, and to test the strength of devices that attach equipment to platforms that can crash.

Gunfire vibration, aircraft

This test assures that equipment mounted in an aircraft with onboard guns can withstand the vibration levels caused by the overpressure pulses emitting from the gun muzzle.

Temperature, humidity, vibration, altitude

This test identifies failures that temperature, humidity, vibration, and altitude can induce in aircraft electronic equipment either individually or in any combination during ground and flight operations. It may be used for other similar purposes. (*Note*: This is the principal test for avionics.)

Icing/freezing rain

This test evaluates the effect of icing produced by a freezing rain, mist, or sea spray on the operational capability of materiel. This method also provides tests for evaluating the effectiveness of deicing equipment and techniques, including field expedients.

Vibro-acoustic, temperature

This test reproduces the combined temperature, vibration, and other operating stresses, as needed, that an externally carried aircraft store will experience during in-service flights.

Appendix

B

Software Documentation Description

This appendix presents a brief description of each document required (optional in some cases) by the major civil and military publications on software development: DO-178, DOD-STD-2167, and DOD-STD-2168.

Part A: DO-178 Documents[1]

Configuration Index document (CID)

The CID is the major control document for a unique software version and serves as an index of all applicable documentation. It provides a historical reference to all documents under configuration control and indicates the current status of each document.

Software requirements document

This document, which may be produced by the installer and/or the equipment manufacturer, should contain (but is not necessarily limited to) the following:

1. The functional and operational requirements of the software, stated in quantitative terms with tolerances where applicable, general and descriptive material including a functional block diagram or equivalent representation of each computer program, graphic illustration of functional operation and the relationship between functions

2. Requirements for each operational function or operating mode, plus special functions such as sequencing, error detection and recovery, input and output control, real-time diagnostics, etc.
3. Performance, test, design, and criticality requirements for each function
4. Sizing and timing requirements
5. Hardware and software interfaces
6. Built-in test and/or monitoring requirements
7. Performance (degradation and function) under fault conditions

Design description document

This document, which is developed by the equipment manufacturer, describes the design of the software program and traceability of requirements from the software requirements document to the design and implementation. It should include a software documentation description that includes the following:

1. A description of the program structure (design trees) and partitioning
2. Data flow description or diagram
3. Program control flow description or diagram
4. Descriptions of data and control interfaces between software partitions and between software and hardware
5. Description of algorithms
6. Timing specification
7. Memory organization and sizing information
8. Program interrupts

It is recognized that as an alternative, some of the information may be contained as comments within the program source listing.

Programmer's manual

The intent of this document is to provide adequate information for understanding and programming the computer used in the airborne equipment. This information should include a complete architecture description (instruction set operation) and programming language description manual. It is recognized that design and interface information contained in the design description document (and others) may be necessary to program the computer.

Software configuration management plan

1. Configuration identification
   a. Specifications
   b. Drawings
   c. Software-peculiar documentation
   d. Numbering systems (documentation, software, and hardware hosting software)
2. Configuration control
   a. Means of establishing formal baselines and of authorizing changes to them
   b. Library controls
   c. Media controls
3. Configuration status accounting
   a. Means of recording and reporting status and configuration of software (and related hardware) elements
4. Review and audits
   a. Means of assuring that the "as-required" configuration is reflected in the "as-designed" configuration, which, in turn, results in the "as-built" configuration
5. Supplier control
   a. The means of application of relevant software configuration management requirements to subtier contractors

Software quality assurance plan (may be combined with the software configuration management plan)

1. Purpose
   a. Relation to product development
2. Quality assurance function
   a. Organization
   b. Work tasks
3. Documentation
   a. Control
   b. Changes
   c. Deliverable and nondeliverable documents subject to configuration control
4. Policies, procedures, and practices
   a. Requirements, design, implementation, test, and documentation
5. Reviews and audits
   a. Scheduled and nonscheduled

6. Configuration management
   a. Means of assuring that adequate procedures and controls are documented and implemented
7. Problem reporting and corrective action
   a. Processing
   b. Tracking
   c. Reporting
8. Media control
   a. Libraries
   b. Protection
9. Testing
   a. Environment
   b. Traceability
10. Supplier control
    a. Assurance that requirements, standards, and controls imposed on the equipment manufacturers are imposed also on the lower-tier manufacturers as applicable
11. Records
    a. Procedures
    b. Traceability

Source listing

This document contains source statements for the computer program annotated to describe modules, functions, and program flow. The listing should include the program part number, the program name, and the date of release and/or version. The linker and locator listing (which provides a map of the load module) should be included in this package.

Source code

This "document" consists of source code in a machine-readable form.

Executable object code

This "document" consists of executable object code in a machine-readable form.

Support/development configuration

This document describes the hardware, software, and processes used to develop and maintain the software and to produce the source and executable code.

Accomplishment summary

This document is considered the primary document for use by regulatory agencies for certification. It should identify all other documents that may be required for information or to be submitted. The document is a *summary,* normally no more than 10 pages long. However, its length will depend on the complexity and criticality of the equipment or system and/or the software level. Particular attention should be given to Item 1 below, including safety, and to Item 5, the summary of test plans and results. The document addresses, as applicable, items such as:

1. Equipment and system description including organization of software
2. Criticality categories and software levels
3. Design disciplines
4. Development phases
5. Software verification plan and results
6. Configuration management
7. Quality assurance
8. Certification plan
9. Organization and identification of documents

Software verification plan, procedures, and results

These documents may be tailored to the various activities defined in Sec. 6.0. They describe tests to be performed, the purpose of each test, coverage analyses, functions to be tested, sequences and methods of testing, and test results. They also cover means of verification, test equipment requirements, software test requirements, and results. If aspects of verification are to be accomplished within the validation process, they should be identified.

Software design standards

This document, which may be prepared by the equipment manufacturer, the installer, or the user, specifies the software design and implementation standards defined as applicable to the software development and test process. It also describes types of software implementations which may jeopardize meeting the functional objectives of the system.

System requirements

The system requirements document should describe the overall system to be certified. It may originate from either the installer or the equipment manufacturer and should contain the following:

1. A system description containing functional block diagrams, line replaceable unit (LRU) component breakdowns, and descriptions of functions to be certified.
2. Certification requirements, including all applicable FARs, JARs, advisory circulars, BCARs, etc. In addition, if the system containing the functions to be certified is an element of a higher-level system, the higher system requirements and criticality should be referenced or summarized.
3. Means of compliance to accomplish the following:
   a. Supplier development programs
   b. Installer validation testing programs
   c. Design error detection and correction program
   d. Documentation
4. Any specific design techniques such as monitoring, redundancy, or functional partitioning.

Plan for software aspects of certification

This document, which is useful at an early stage, may or may not form part of the certification plan. It should include the following and may be satisfied by a preliminary issue of the accomplishment summary:

1. Brief equipment and system description, as appropriate
2. Criticality categories and software levels
3. Activities in support of software considerations for certification
4. Documentation plan
5. Schedule
6. Organizations—their involvement and responsibilities

Part B: DOD-STD-2167 Documents[2]

System/segment design document

The system or segment design document (SSDD) describes the design of a system or segment and its operational and support environments. It describes the organization of a system or segment as composed of hardware configuration items (HWCIs), computer software configuration items (CSCIs), and manual operations.

Software development plan

The software development plan (SDP) describes a contractor's plans for conducting software development. The SDP is used to give the government insight into the organization(s) responsible for performing software development and the methods.

Software requirements specification

The software requirements specification (SRS) specifies the engineering and qualification requirements for a computer software configuration item (CSCI). The SRS is used by the contractor as the basis for the design and formal testing of a CSCI.

Interface requirements specification

The interface requirements specification (IRS) specifies the requirements for one or more interfaces between one or more computer software configuration items (CSCIs) and other configuration items or critical items.

Interface design document

The interface design document (IDD) specifies the detailed design for one or more interfaces between one or more computer software configuration items (CSCIs) and other configuration items or critical items.

Software design document

The software design document (SDD) describes the complete design of a computer software configuration item (CSCI). It describes the CSCI as composed of computer software components (CSCs) and computer software units (CSUs).

Software product specification

The software product specification (SPS) consists of the software design document (SDD) and source code listings for a computer software configuration item (CSCI). Upon government approval and authentication following the physical configuration audit (PCA), the SPS establishes the product baseline for the CSCI.

Version description document

The version description document (VDD) identifies and describes a version of a computer software configuration (CSCI). The VDD is used by the contractor to release CSCI versions to the government. The term *version* may be applied to the initial release of a CSCI.

Software test plan

The software test plan (STP) describes the formal qualification test plan for one or more computer software configuration items (CSCIs). The STP identifies the software test environment resources required for formal qualification testing (FQT) and provides schedules for FQT activities. In addition, the STP identifies the individual tests to be performed during FQT.

Software test description

The software test description (STD) contains the test cases and test procedures to perform formal qualification testing of a computer software configuration item (CSCI) identified in the software test plan (STP).

Software test report

The software test report (STR) is a record of the formal qualification testing performed on a computer software configuration item (CSCI). The STR provides the government with a permanent record of the formal qualification testing performed on a CSCI.

Computer system operator's manual

The computer systems operator's manual (CSOM) provides information and detailed procedures for initiating, operating, maintaining, and shutting down a computer system and for identifying and/or isolating a malfunctioning component in a system. A CSCM is developed for each computer system in which one or more CSCIs execute.

Software user's manual

The software user's manual (SUM) provides user personnel with instructions sufficient to execute one or more related computer software configuration items (CSCIs). The SUM provides the steps for executing the software, the expected output, and the measures to be taken if error messages appear.

Software programmer's manual

The software programmer's manual (SPM) provides information needed by a programmer to understand the instruction set architecture of the specified host and target computers. The SPM provides information that may be used to interpret, check out, troubleshoot, or modify existing software on the host and target computers.

Firmware support manual

The firmware support manual (FSM) provides the information necessary to load software or data into firmware components of a system. It is equally applicable to read-only memory (ROMs), programmable ROMs (PROMs), erasable PROMs (EPROMs), and other firmware devices.

Computer resources integrated support document

The computer resources integrated support document (CRISD) provides the information needed to plan for life cycle support of deliverable software. The CRISD documents the contractor's plans for transitioning support of deliverable software to the support agency.

Part C: DOD-STD-2168 Document[3]

Software quality program plan

The software quality program plan (SQPP) identifies the organizations and procedures to be used by the contractor to perform activities related to the software quality program specified by DOD-STD-2168. The SQPP is used to evaluate the contractor's plans for implementing the software quality program.

References

1. DO-178A Software Considerations in Airborne Equipment Certification, RTCA, 1985.
2. DOD-STD-2167A Defense System Software Development, 29 February 1988.
3. DOD-STD-2168 Defense System Software Quality Program, 29 April 1988.

Appendix C

Document Ordering Addresses

ARINC Characteristics, Reports
Aeronautical Radio, Inc.
2551 Riva Road
Annapolis, Maryland 21401

National Standards
American National Standards
 Institute
11 West 42d Street
New York, New York 10036

FAA Aviation Regulations,
 Advisory Circulars
Superintendent of Documents
710 North Capitol Street
Washington, DC 20402

NASA documents and DOD
 documents for private citizens
National Technical Information
 Service
5285 Port Royal Road
Springfield, Virginia 22161

NRC regulations
Nuclear Regulatory Commission
Document Sales Division
1717 H Street, N. W.
Washington, DC 20555

SAE documents
SAE
400 Commonwealth Drive
Warrendale, Pennsylvania 15096

AIAA documents
American Institute of Astronautics
 and Aeronautics
Technical Information Service
555 West 47th Street
New York, New York 10019

DOD documents (contractors only)
Defense Technical Information
 Center
Cameron Station
Arlington, Virginia 22314

IEEE documents
Institute of Electrical and
 Electronics Engineers
445 Hoes Lane
Piscataway, New Jersey 08854

DOD standards and specifications
Naval Publications and Forms
 Center
5801 Tabor Avenue
Philadelphia, Pennsylvania 19120

RTCA documents
RTCA
1140 Connecticut Avenue, N. W.
Suite 1020
Washington, DC 20036

Index

A-10 aircraft, 2
A-320 aircraft, 110, 131–133, 134, 135
Abbreviated Test Language for Avionics Systems (ATLAS), 103, 111, 112-113
Abstract Semi-Markov Specification Interface to the SURE Tool (ASSIST), 186–187
Accessibility, 107, 187
Acquisition cost, 227, 230
Ada, 128, 211–216
Adaptability, 188
Adjustments, tax, 222, 226
Advisory Circular (*see* Federal Aviation Administration)
Aeronautical Radio, Inc.:
　ARINC 404, 140
　ARINC 429, 31–37, 38
　ARINC 600, 139–142, 145
　ARINC 604, 109
　ARINC 629, 37, 39–43
　ARINC 659, 43–46
Affordability, 188
AIMS (Airplane Information Management System), 128–131
Air cooling, 141–142, 143–144, 145
Aircraft:
　F-16, 121–122, 123
　F-22, 29, 144–145
　interfaces, 148–149
Airplane Information Management System (AIMS), 128–131
Algebra, boolean, 161–162
Alphanumeric words, 33, 34–35
Analysis:
　cash flow, 226–227, 228–230
　function hazard, 157
Analytical redundancy, 100
ANSI/IEEE Standard 416, 112
ANSI/MIL-STD-1815 (*see* Ada)
Architectures:
　centralized, 119–120
　cost of evaluating, 153
　distributed, 120

Architectures (*Cont.*):
　evaluation of, 152–154, 187–188
　examples of, 121–137
　federated, 120
　instruction set (*see* MIL-STD-1750)
ARINC (*see* Aeronautical Radio, Inc.)
Arrangement, instruments in T, 70
ASSIST (Abstract Semi-Markov Specification Interface to the SURE Tool), 186–187
ATE (Automatic test equipment), 111–112
ATLAS (*see* Abbreviated Test Language for Avionics Systems)
Automatic test equipment (ATE), 111–112
Availability, 153, 187

B-757 aircraft, 70
B-767 aircraft, 70
B-777 aircraft, 128
Battery, 81, 88, 89
Beechcraft Starship aircraft, 133, 135–137
Benefits, operating, 226, 227
BITE (built-in test equipment), 107–109
Blocks, recovery, 97
Bonding, 183
Boolean algebra, 161–162
Built-in test equipment (BITE), 107–109
Bus (*see* Data bus)
Bus controller, 20, 27
Bus monitor, 20

CA (*see* Criticality analysis)
CAA (Civil Aviation Authority), 153n
Capability, 152
CARE III (Computer-Aided Reliability Estimation, Version III), 184–185
Cash flow analysis, 226–227, 228–230
Cathode ray tube (CRT), 50–54
　color, 52–54
　monochrome, 50–51
　multifunction keyboard, 61

273

Index

Central fault display system (CFDS), 109–111
Certification, 154–157
Certifiability, 153
CFDS (central fault display system), 109–111
Civil Aviation Authority (CAA), 153n
Cockpit:
  controls, 69
  design, 67–71
COCOMO (constructive cost model), 232–234, 235–238
Coding, 192, 194
Command word, 21–22
Compatibility, 188
Computer-Aided Reliability Estimation, Version III (CARE), 184–185
Constructive cost model, 232–234, 235–238
Controller, bus, 20, 27
Controls, cockpit, 69
Cooling, 145–148
  air, 141–142, 143–144, 145
Cost:
  evaluating architectures, 153
  of ownership, 222
  of retirement, 223, 226
Costs, operating, 223, 225–226
Criticality analysis (CA), 169–172
  classification, 164
  number, 170–172
CRT (*see* Cathode ray tube)

Damage modes and effects analysis (DMEA), 172–173
Data bus:
  ARINC 429, 31–37, 38
  ARINC 629, 37, 39–43
  ARINC 659, 43–46
  DOD-STD-1773, 27–29
  HSDB, 29–31
  MIL-STD-1553, 20–27
  MIL-F-9490, 16
  optical (*see* Optical data bus)
  star, 27–28
Data word:
  ARINC 429, 32–35
  ARINC 629, 39–41
  MIL-STD-1553, 24
DATAC (Digital Autonomous Terminal Access Communication), 41
Decision/action diagram, 8

Design:
  of cockpits, 67–71
  of power system, 85, 88–90
Design evaluation matrix, 11–12
DFCS (digital flight control system), 121–123
DGAC (Direction Generale de l'Aviation Civile), 153n
Diagrams:
  decision/action, 8
  functional flow, 8–9
Digital Autonomous Terminal Access Communication (DATAC), 41
Digital flight control system (DFCS), 121–123
Direction Generale de l'Aviation Civile, 153n
Discrete words:
  ARINC 429, 33–34
  ARINC 629, 40
Displays:
  head level (HLD), 64–65
  head up (HUD), 61–64
  helmet-mounted (HMD), 65–66
  raster, 53–54
  stroke, 53, 54
  (*See also* Cathode ray tube; Flat panel displays)
DMEA (damage modes and effects analysis), 172–173
DO-160:
  electromagnetic interference (EMI), 180–181
  environmental testing, 173–176
  interfaces, 148
  power quality, 81–85, 86–88
DO-178, 200–205
DOD-HDBK-763, 6–11, 12
DOD-STD-1773, 27–29
DOD-STD-1788, 142–144, 145
DOD-STD-2167, 205–211
DOD-STD-2168, 211

EFA (Eurofighter Aircraft), 127–128
Efficiency, luminous, 75
EL (electroluminescent), 55
Electromagnetic interference (EMI), 178–184
Energy, luminous, 75
Environmental testing:
  DO-160, 173–176, 249–258
  MIL-STD-810, 176–178, 258–260
Eurofighter Aircraft (EFA), 127–128

Evaluation of architectures, 152–154, 187–188
Extremely improbable, 155–156

F-16 aircraft, 121–122, 123
F-22 aircraft, 29, 105, 144–145
FAA (*see* Federal Aviation Administration)
Failure modes and effects analysis, 162–169
FAR (*see* Federal Aviation Regulation)
Fault detection, 92
Fault-tolerant hardware:
  duplication and comparison, 98
  self-checking, 98–99
  triplication and voting, 96–97
Fault-tolerant software, 99–100
Fault tree analysis (FTA), 157–162
  certification, 157–158
Federal Aviation Administration (FAA), Advisory Circular:
  25-11, 59
  25.1309-1A, 105, 155–157
  21-16, 174
Federal Aviation Regulation (FAR):
  25.581, 182
  25.671, 154
  25.672, 154
  25.1303, 69, 155
  25.1309, 14, 91
  25.1321, 70
  25.1322, 70
FHA (functional hazard analysis), 157
FIC (fleet installed cost), 223–224
Filtering, 183
Fitts list, 8–9
Flat panel displays:
  electroluminescent, 55
  gas discharge, 55–56
  light emitting diode (LED), 55
  liquid crystal:
    backlighted, 56–57
    reflective, 56
  luminous, 54–56
Fleet installed cost (FIC), 223–224
Fleet spares cost (FSPC), 223, 224–225
Fleet support cost (FSUC), 223, 225
Flexibility, 188
Flight critical requirements, 6
Flight phase essential, 14
FMEA (Failure modes and effects analysis), 162–169

FSPC (fleet spares cost), 223, 224–225
FSUC (fleet support cost), 223, 225
FTA (*see* Fault tree analysis)
Function:
  allocation, 8–10
  assignment, 8–10
Functional flow diagrams, 8–9
Functional hazard analysis, 157

Generation:
  raster image, 53–54
  stroke image, 53, 54
Grounding, 183

Hardened kernels, 100
Hardware, fault-tolerant (*see* Fault-tolerant hardware)
Head level display (HLD), 64–65
Head-up display (HUD), 61–64
Heat pipes, 147–148
Helmet-mounted display (HMD), 65–66
HLD (head level display), 64–65
HMD (helmet mounted display), 65–66
HUD (head-up display), 61–64

Ilities, 152–154, 187–189
Improbable, 155–156
Instruction set architecture (*see* MIL-STD-1750)
Instruments, T arrangement of, 70
Interchangeability, 188
Interfaces, aircraft, 148–149
Investment cost, 222–224
Invulnerability, MIL-F-9490, 15–16

Keyboard, multifunction, 60–61
Kit protection level, 238–240

LCC (*see* Life cycle cost)
LCD (liquid crystal display), 56–57
LED (light emitting diode), 55
Life cycle cost (LCC), 221–231
  cash flow analysis, 226–230
  civil, 223–226
  effect of cooling, 147
  evaluating architectures, 152, 153
  military, 227, 230–231
Light emitting diode (LED), 55
Lightning:
  electromagnetic interference, 179, 181–182
  DO-160, 181–182
  MIL-F-9490, 16

276 Index

Line replaceable unit (LRU)
  ARINC 600, 139–142, 145
  DOD-STD-1788, 142–145
List, Fitts, 8–9
LRU (see Line replaceable unit)
Lumen, 75
Luminance, 75
Luminous efficiency, 75
Luminous energy, 75
Luminous flat panels, 54–56
Luminosity curve, 75–76

MAFT (Multiprocessor architecture for fault tolerance), 93–95
Maintainability, evaluating architectures for, 153
Maintenance, 104–107
  location, 105
  personnel, 104
  time, 104–105
Manchester coding:
  DOD-STD-1773, 27
  MIL-STD-1553, 21
MCU (modular concept unit), 144
Mean time between failures (MTBF), 114–117
Mean time between unconfirmed removals (MTBUR), 5
MEL (minimum equipment list), 169
Methods, structured, 195–200, 201
MFK (multifunction keyboard), 60–61
MIL-B-5087, 183
MIL-E-5400, 127
MIL-E-6051, 80, 127, 179–180
MIL-F-9490, 13–17, 81
MIL-HDBK-217, 114–115
MIL-HDBK-253, 183–184
MIL-M-28787, 126, 144–145
MIL-M-38510, 126
MIL-STD-203, 69
MIL-STD-461, 181
MIL-STD-462, 180
MIL-STD-463, 181
MIL-STD-470, 105
MIL-STD-704, 78–81, 86–88
MIL-STD-721, 114
MIL-STD-785, 147
MIL-STD-810:
  environmental testing, 176–178
  interfaces, 148
MIL-STD-882, 164
MIL-STD-1553, 20–27, 127
  MIL-F-9490, 16

MIL-STD-1629, 163–164, 169–172
MIL-STD-1750, 126, 216–218
MIL-STD-2069, 172–173, 174
MIL-STD-2165, 108–109
Minimum acceptable rate of return, 227, 228, 229
Minimum cut set, 162
Minimum equipment list (MEL), 169
Mission requirements, 2
  profile, 6
  scenario, 6
Mode code, 22–23
Modular concept unit (MCU), 144
Money, time value of, 226–227
Monitor, bus, 20
MTBF (mean time between failures), 114–117
MTBUR (mean time between unconfirmed removals), 5
Multifunction keyboard, 60–61
Multiprocessor architecture for fault tolerance (MAFT), 93–95
Multiversion software, 99

N-version software, 99
Nominal programmer months, 232–234
NUREG 0492, 158–162

Operating benefits, 226, 227
Operating costs, 223, 225–226
Operational state, 14
Optical data bus:
  ARINC 629, 42–43
  DOD-STD-1773, 26–29
  MIL-F-9490, 13–17
Ownership, cost of, 222

Partitioning, 120–121
Pave Pillar, 122, 124–127
  requirements, 2
Payback period, 227
Pipes, heat, 147–148
Power, 77–90
  evaluating architectures for, 154
Power system design, 85, 88–90
Probable, 156
Programmability, 188
Programmer months, nominal, 232–234
Protection level, kit, 238–239

Raster image generation, 53–54
Reconfiguration, 92
Recovery blocks, 97

Redundancy:
  analytical, 100
  hardware, 91
  software, 91
Reliability, 114–117, 152–153
Remote terminal (RT), 21, 27
Repairability, 187
Requirements, 2
  flight critical, 6
Retirement cost, 223, 226
Return rate, minimum, 227, 228, 229
Risk, technical, 154
RT (remote terminal), 21, 27

SAFRA (Semi-Automated Functional Requirements Analysis), 11, 13
Semi-Markov Unreliability Range Estimator (SURE), 184–185
Shielding, 183
Software:
  cost, 232–234, 235–238
  development, 192–195
  DO-178, 200–205
  documentation description, 261–269
  DOD-STD-2167, 205–211
  DOD-STD-2168, 211
  fault-tolerant, 99-100
  multiversion, 99
  N-version, 99
  partitioning, 193
  redundancy, 91
  test plan, 194
Spares level, 234, 238–240
Speech:
  recognition, 71–73, 74
  synthesis, 71, 73, 75
State, operational, 14
Status word:
  ARINC 629, 39, 41
  MIL-STD-1553, 23–24
Stroke image generation, 53, 54
Structured methods, 195–200, 201
Supportability, 188
SURE (Semi-Markov Unreliability Estimator), 184–185

Survivability, 188
  evaluating architectures for, 153–154
  MIL-F-9490, 15
  MIL-STD-2069, 172–173
Susceptibility, 188
  MIL-STD-2069, 172–173

T (instruments arrangement), 70
Tax adjustments, 222, 226
Technical risk, 154
Test equipment (automatic test equipment), 111–112
Testability, MIL-STD-2165, 108–109
Testing:
  built-in test equipment, 107–109
  environmental (see Environmental testing)
  line replaceable unit (LRU), 109
Time:
  in mission phase, 5
  warm up, MIL-F-9490, 17
Time value of money, 226–227
Timelines, 68
Transients, 80

Unit cost, 224

Verification, 202–203
Voice interactive systems, 71–75
  recognition, 72–73, 74
  synthesis, 74–75
Vulnerability, 188
  MIL-STD-2069, 172–173

Weight:
  ARINC 600, 142–145
  DOD-STD-1788, 143, 145
  evaluating architectures, 154
Word, command, 21–22
Words:
  alphanumeric, 33, 34–35
  discrete (see Discrete words)
  status (see Status word)
Workload analysis, 68–69

ABOUT THE AUTHOR

Cary R. Spitzer is a manager in the Advanced Transport Operating Systems Program Office at NASA where he recently led a flight test program that culminated in the first satellite-guided landing of a commercial airliner. He was nominated for the 1991 Collier Trophy in recognition of this work. He is a former deputy manager of NASA's Mars Viking program. Mr. Spitzer is a Fellow of the IEEE and an Associate Fellow of the AIAA and has played a major role in organizing the IEEE-AIAA conferences on digital avionics systems.